Green Engineering Materials

The use of green engineering materials represents a promising approach to sustainable development. This book explores various renewable materials and their properties, applications, and technological advancements driving their use in modern engineering and construction.

This book highlights the significance of green engineering materials in sustainable development and explains their effective use in engineering and construction. It covers bamboo as a rapidly growing renewable material in use with significant engineering potential, detailing its unique characteristics, preservation methods, and uses in construction. The book also investigates sustainable plant-based composites, focusing on biopolymer and biomass matrices, cellulose-based materials, lignin, polylactic acid, and natural rubber. It highlights the benefits of plant fibres like rice husk ash and jute while addressing the challenges in adopting these composites in engineering. Green concrete technologies like hybrid geopolymers and green additives and their manufacturing processes are also discussed. Following this, the book discusses bio-based adhesives and then explores the industrial potential of rice husk ash in applications including electronic devices, composites, and reinforced concrete. Palm oil and coconut shells are also examined as versatile renewable resources for various construction and development applications. Finally, the book emphasizes the importance of wood in construction, including its properties, treatment methods, and future trends in sustainable construction practices.

Because this has a broad scope and provides readers with a basic level of technical knowledge, it is an ideal guide for general readers concerned with sustainability as well as anyone starting out in the field, including undergraduate students and readers in the industry who want to keep abreast of current developments and trends in this field.

Yuli Panca Asmara is an Associate Professor at INTI International University. He obtained his BEng in 1995 from Sepuluh Nopember Institute of Technology, Indonesia; his MS in 2002 from the University of Manchester, UK, and PhD in 2010 from University Technology Petronas, Malaysia. He also provides consultancy services on materials and corrosion issues for local companies, petroleum industry training, and supports the turnaround of Abu Dhabi National Oil Company (ADNOC) Liquefied Natural Gas facilities.

Green Engineering Materials

Innovations and Applications for Sustainable Construction

Yuli Panca Asmara

CRC Press
Taylor & Francis Group
Boca Raton London New York

CRC Press is an imprint of the
Taylor & Francis Group, an **informa** business

First edition published 2025
by CRC Press
2385 NW Executive Center Drive, Suite 320, Boca Raton FL 33431

and by CRC Press
4 Park Square, Milton Park, Abingdon, Oxon, OX14 4RN
CRC Press is an imprint of Taylor & Francis Group, LLC

ISBN: 9781041020639 (hbk)
ISBN: 9781041020646 (pbk)
ISBN: 9781003617594 (ebk)

DOI: 10.1201/9781003617594

Typeset in Palatino
by Deanta Global Publishing Services, Chennai, India

Contents

Preface

Green Engineering Materials is a comprehensive exploration of how innovative, eco-friendly materials are shaping the future of industries worldwide. This book delves into the science, application, and potential of materials designed to minimize environmental impact while enhancing performance and functionality.

Engineering and material science are undergoing a transformative period, driven by the urgent need to address environmental challenges such as climate change, resource depletion, and pollution. Traditional materials, though reliable and widely used, often come with a high environmental cost. This transition towards sustainable alternatives, ranging from plant-based composites and renewable resources like bamboo and coconut shells to advanced materials like green concrete and bio-based adhesives, represents a significant step forward in achieving a sustainable balance between development and environmental stewardship.

Each chapter in this book provides an in-depth analysis of specific materials and technologies. Starting with a broad introduction to green engineering materials, the book progresses through specialized topics, such as the use of bamboo, plant-based composites, and innovative solutions like rice husk ash and palm oil derivatives. It also explores emerging opportunities for green engineering in construction, infrastructure, and beyond. The topics discussed are not limited to technical attributes; they also include the economic, environmental, and social implications of adopting these sustainable alternatives. Challenges in implementation and market acceptance are addressed alongside case studies and practical applications, offering a balanced view of the opportunities and hurdles in this evolving field.

Green Engineering Materials serves as a valuable resource for engineers, scientists, students, and industry professionals committed to fostering innovation while prioritizing environmental responsibility. It seeks to inspire readers to embrace the transition toward sustainable practices, harnessing the potential of green materials to shape a better, more sustainable future.

Through this book, we aim to spark further research, collaboration, and development, paving the way for the next generation of engineering materials that combine innovation with sustainability.

Nilai, November 2024
Yuli Panca Asmara

About the Author

Yuli Panca Asmara is an Associate Professor at INTI International University. He obtained his BEng in 1995 from Sepuluh Nopember Institute of Technology, Indonesia; MS in 2002 from the University of Manchester, UK, and PhD in 2010 from University Technology Petronas, Malaysia. He provides consultancy services on materials and corrosion issues for local companies, petroleum industry training, and supports the turnaround of Abu Dhabi National Oil Company (ADNOC) Liquefied Natural Gas facilities.

Acknowledgements

This book has been compiled with the assistance of many parties. We would like to express our gratitude to Prof. Wong Ling Shing, Dr Chan Siew Chong, and Dr Keng Hoo Chuah, as well as all the staff at FEQS, Inti International University, Malaysia, for their support that made this book possible. We would also like to thank our family members:

Nunuk Ariyani, Ferro Handaru Adidarma, and Pascal Adiwidya Adiluhur.

1

Green Engineering Materials: Combining Innovation and Sustainability for a Better Future

1.1 Engineering Materials and Materials Selections

Engineering materials are the backbone of modern industry, providing the essential building blocks for manufacturing, construction, and technological advancement. These materials are categorized broadly into metals, polymers, ceramics, and composites, each offering unique properties suited to specific applications [1, 2]. Metals, including steel, aluminium, and copper, are widely used for their strength, ductility, and conductivity. Metals and alloys are renowned for their exceptional tensile strength and durability, making them indispensable in construction, heavy machinery, and infrastructure. Polymers, or plastics, are versatile engineering materials that offer lightweight and flexible alternatives to metals. These materials, such as polythene, polypropylene, and polytetrafluoroethylene (PTFE), are often chosen for their resistance to corrosion, chemical stability, and ease of manufacturing [3, 4]. Ceramics, a category that includes materials like alumina, silicon carbide, and zirconia, are valued for their hardness, high-temperature stability, and resistance to wear. Composites, combining two or more different materials, offer a synergy of properties that individual components alone cannot provide. Common examples include carbon fibre-reinforced polymers (CFRP) and glass fibre-reinforced polymers (GFRP). Advanced materials, such as nanomaterials and biomaterials, represent the forefront of engineering materials research [5, 6]. Nanomaterials, which are engineered at the atomic or molecular scale, exhibit unique properties like enhanced strength, conductivity, and reactivity, making them ideal for applications in electronics, energy storage, and medicine [7].

Metals, polymers, ceramics, and composites have each seen significant advancements in development [8, 9]. However, the mining of metals leaves behind excavation pits, and the continuous extraction of these materials can lead to the depletion of resources and environmental degradation. Over

DOI: 10.1201/9781003617594-1

time, the cumulative impact of mining can limit the availability of high-quality ores, disrupt ecosystems, and contribute to soil and water pollution. Additionally, the processes of fabrication, refining, and transportation are complex and often generate considerable greenhouse gas emissions. From raw material extraction to final manufacturing, producing these materials requires substantial energy, mostly sourced from fossil fuels, which itself is a source of ongoing controversy due to its environmental impact [10, 11].

The availability of various material options makes material selection a critical aspect of designing and constructing structures that meet engineering requirements. Several factors must be considered to ensure the structural integrity of the materials chosen. Key properties include hardness, strength, stability at high temperatures, corrosion resistance, and resilience [12, 13]. Environmental conditions such as temperature, humidity, pH levels, and exposure to corrosive substances should also be carefully evaluated. Compatibility with other materials and components is crucial, as incompatibility can lead to issues like galvanic corrosion. Additionally, maintenance requirements and cost considerations play a significant role. Regular inspections, protective coatings, and specialized treatments are often necessary to maintain and enhance corrosion resistance [14, 15].

The availability and cost of engineering materials should align with project budgets while considering long-term maintenance expenses. Sustainable materials are increasingly becoming a preferred choice due to their environmental benefits, including reduced pollution and enhanced sustainability. With the integration of innovative solutions and advanced technologies, *green engineering materials* have demonstrated significant potential to meet the ever-evolving demands of industries operating under complex conditions and requirements, while simultaneously achieving environmental sustainability goals.

1.2 The Evolution of Green Engineering Materials

Environmental concerns have accelerated the development of sustainable engineering materials such as biodegradable polymers, recycled metals, and bio-based composites. These materials are critical as industries strive to reduce waste, lower carbon footprints, and conserve resources. Green engineering materials have been evolving since the earliest civilizations, transforming into a modern movement driven by technology and innovation to reduce the environmental impact [16–18]. The evolution of green construction demonstrates a clear trajectory of growth and adaptation over time. The publication trends on Google Scholar for the three key green engineering materials—rice husk, bamboo, and palm oil fibre—reveal significant growth in research interest over five decades. From 1970 to 1980, there were 2,860 papers on rice husk, 15,700 on bamboo, and 1,400 on palm oil fibre, highlighting bamboo's early dominance in research. This trend continued through the 1980s and 1990s, with bamboo leading in publication counts, reaching 20,600 papers in 1980–1990 and 44,800 in 1990–2000, compared to rice husk's 6,060 and 11,400, and palm oil fibre's 2,880 and 6,180. Between 2000 and 2010, the number of studies increased drastically, with bamboo seeing 180,000 publications, rice husk reaching 17,500, and palm oil fibre rising to 15,900. The 2010–2020 period marked the peak of research activity, with bamboo amassing 333,000 papers, followed by rice husk with 91,000, and palm oil fibre with 24,300 [18–20]. These trends underscore the increasing importance of these natural materials in sustainable engineering, with bamboo consistently leading as a versatile and widely studied green engineering material.

For thousands of years, various forms and techniques in green materials have existed, often characterized by the use of natural, locally sourced materials like straw bales—a practice that originated in ancient civilizations and remains in use today. However, it wasn't until the 20th century that sustainable building practices gained widespread popularity, as architects and builders became more aware of the environmental impact of buildings. Passive solar technology, which emerged prominently in the 1970s, is an example of this shift; it harnesses the sun's natural heat and light for building temperature control, reducing reliance on artificial heating and cooling systems. In the 1990s, the US Green Building Council (USGBC) was established and introduced the Leadership in Energy and Environmental Design (LEED) certification, setting a global benchmark for sustainable construction [21, 22]. As more organizations adopted these standards, green building principles began spreading worldwide, with notable environmental benefits and cost savings from durable materials and reduced energy consumption.

Today, as demand for construction materials surges, the green building industry continues to expand, spurred by new technologies and innovative practices. Green roofs, which help manage and filter rainwater, are increasingly popular in urban areas, while sustainable materials like laminated timber are becoming preferred alternatives to concrete and steel. Smart technologies are also now integral to green buildings, with IoT-enabled systems and energy-efficient designs optimizing energy use on a global scale. This ongoing evolution reflects the potential of green construction to meet growing needs sustainably, shaping a resilient future for the industry [23].

1.3 Environmental Benefits of Green Engineering Materials

Green engineering materials offer significant environmental benefits by reducing resource depletion, minimizing pollution, and supporting a low-carbon economy. Unlike traditional materials that require energy-intensive processes, green materials are often renewable, recyclable, or biodegradable, aligning with eco-friendly practices. For instance, bamboo, used in projects like the *Green School Bali*, replaces carbon-intensive construction materials while sequestering CO_2 during growth. Similarly, India's plastic roads repurpose waste, reducing landfill burden and enhancing road durability. Hempcrete, a biocomposite made from hemp fibres, provides superior insulation and energy efficiency, as seen in French housing projects where energy use dropped by 40% [24]. These examples illustrate how green engineering materials not only mitigate environmental impact but also drive innovation, paving the way for sustainable development across industries.

Green technology plays a critical role in reducing CO_2 emissions, especially given the significant contribution of fossil fuel combustion to global warming since the Industrial Revolution. Innovations in renewable energy and energy efficiency could account for over 60% of the necessary emissions reductions, according to the International Energy Agency (IEA). However, the effectiveness of these technologies depends on regional factors like urbanization, industrialization, and economic structures. While green innovations are essential for sustainability, their actual impact on emissions remains debated and requires further research [25].

In 2019, the EU's cement, steel, and chemical industries incurred significant external costs, amounting to €84 billion, €202 billion, and €169 billion, respectively. These costs primarily stem from their contributions to climate change, fossil fuel depletion, and air pollution. The steel industry, particularly due to air pollution (including particulate matter and human toxicity), had the largest share of damages. External costs vary by country, with Germany, France, Italy, and Spain, being the highest contributors. On average, the external costs per euro of product in the EU 27 for these industries were €0.84 for cement, €2.01 for steel, and €2.52 for chemicals [26].

Efforts to improve the sustainability of Ordinary Portland Cement (OPC) have led to exploring alternatives with lower carbon footprints. The calcination process in OPC production is a major environmental concern, incurring external costs of €222.08 per tonne from climate change and €83.03 per tonne from air pollution in Italy (2019). Eco-friendly alternatives priced below €400 per tonne could be more economical if external costs are negligible. Green cement, such as geopolymer cement made from fly ash, offers a sustainable solution. With external costs of €38.85 per tonne for climate impacts and €0.10 for air pollution, geopolymer cement reduces environmental damage by repurposing waste and optimizing energy use. Producing steel in Germany using the blast furnace method in 2019 resulted in external costs of €618.20 per tonne due to climate change and €823.91 per tonne due to air pollution. In comparison, the average price of hot rolled steel coil in Europe was €470 per tonne. This indicates that an environment-friendly steel production method with a market price under €1,900 per tonne would still be more cost-effective if its environmental impact is negligible [27].

Some companies are already exploring sustainable innovations, such as replacing conventional blast furnaces with green hydrogen-based reduction processes, as seen in the HYBRIT pilot project. This alternative method reduces external costs significantly to €120.61 for climate impacts and €96.74 for air pollution per tonne of steel. This demonstrates the steel industry's potential to reduce its environmental footprint while meeting society's demand for sustainable materials [28, 29].

Chemical products, including plastics, are essential to many industries and consumers. However, traditional plastic production relies heavily on fossil fuels, harming the environment. For example, producing one tonne of high-density polythene (HDPE) pipes using conventional steam cracking in France in 2019 resulted in €455.10 in climate-related external costs and €180.02 from air pollution. In contrast, electric steam cracking powered by green electricity reduced these costs to €18.84 and €0.87 per tonne, respectively. The average market price of HDPE pipes in 2019 was €1,260 per tonne. Considering environmental costs, green alternatives priced below €1,890 per tonne would be economically competitive. Fossil fuels in plastics production are used as feedstock and fuel, but alternatives exist. Transitioning to electric steam crackers and sustainable feedstock could help the chemical industry reduce emissions and align with EU climate goals, emphasizing the need for sustainable practices and alternative materials [30].

1.4 The Future of Green Engineering: Materials for a Better Tomorrow

Innovative approaches in sustainable material development offer a wide range of opportunities to reduce environmental impact, lower costs, and improve the longevity of structures. While many of these materials come with higher upfront costs, their long-term benefits—including reduced resource consumption, energy savings, and extended service life—make them increasingly attractive to industries worldwide [16, 17]. As technology advances and market demand for sustainable solutions grow, the costs associated with these materials will continue to decrease, paving the way for a greener and more sustainable future in engineering and construction.

The future of green engineering lies in the development and adoption of materials that balance performance, sustainability, and cost-effectiveness. Advanced research is paving the way for materials that are not only durable and efficient but also environmentally friendly. Innovations such as biodegradable plastics, bio-based composites, and self-healing materials are set to revolutionize industries ranging from construction to electronics [31, 32]. For example, 3D-printed structures using recycled materials have proven both sustainable and versatile, while graphene—a material derived from carbon—offers incredible strength and conductivity with minimal environmental impact. The integration of these materials into mainstream applications holds the promise of reducing carbon footprints, conserving natural resources, and fostering a circular economy [33, 34]. By embracing green engineering materials, industries can contribute to a cleaner, more sustainable future for generations to come.

From bio-based polymers to recycled composites, these materials are designed to minimize environmental impact while meeting high-performance standards [35]. For instance, the construction industry is increasingly adopting materials like hempcrete, a carbon-negative alternative to concrete, and cross-laminated timber, which sequester carbon while providing exceptional strength [36]. Similarly, in packaging, biodegradable plastics derived from plant-based sources like cornstarch are reducing reliance on fossil fuels. The automotive sector is also leveraging lightweight materials such as aluminium alloys and bio-composites to improve fuel efficiency and reduce emissions [37]. These advancements demonstrate the potential of eco-friendly innovations to revolutionize engineering practices, paving the way for a more sustainable and resilient future.

1.5 Innovative Green Solutions for CO_2 Mitigation

Human activities, particularly the burning of coal and oil, are the primary contributors to global warming, with industrial and economic growth intensifying both energy consumption and air pollution worldwide. While improving productivity increases the quality of human life, it has also led to severe environmental consequences. According to the *World Energy Outlook 2017*, global energy-related CO_2 emissions are projected to increase slightly by 2040, highlighting the inadequacy of current policies to avert significant climate impacts. This underscores the urgent need for effective strategies, including the adoption of green technologies such as renewable energy and energy efficiency systems [10].

The International Energy Agency (IEA) estimates that green technology could reduce over 60% of CO_2 emissions globally. However, the actual impact of such innovations varies significantly based on socioeconomic conditions. Factors like prosperity, industrial structure, international trade, urbanization, and energy frameworks influence the relationship between human activity, green innovation, and CO_2 emissions. For instance, studies have indicated that urbanization has contributed to CO_2 reductions in China, while free trade has been found to increase SO_2 emissions but reduce them through urbanization in developing Asian countries [10, 38, 39].

Despite the potential of green innovation, its effectiveness in mitigating CO_2 emissions remains debated. Studies have observed that the impact of green innovation on achieving climate goals is limited while existing green technologies have yet to significantly reduce emissions. In contrast, other research suggests that while green innovation enhances environmental productivity, its direct effect on reducing CO_2 emissions remains minimal [40]. This complex interplay of factors implies that while green technologies hold promise, their development must be aligned with broader economic and social strategies to achieve meaningful climate outcomes.

1.6 Technical Challenges in Adopting Green Engineering Materials

Adopting green engineering materials presents numerous technical challenges that need to be addressed for their widespread use in various industries. From ensuring the performance of these materials under extreme conditions to making them compatible with existing manufacturing processes and ensuring scalability, these obstacles must be overcome through continued innovation, research, and investment in advanced manufacturing technologies. While progress is being made, overcoming these technical hurdles is essential to realizing the full potential of green engineering materials in the transition to a more sustainable future.

Substituting existing technologies and engineering materials with green engineering materials faces several technical challenges that must be addressed for their widespread implementation. First, performance under extreme conditions, such as high pressure, temperature, or corrosive environments, remains a concern. For instance, bio-based composites may struggle to maintain strength in deep-sea drilling or under high thermal stress in industrial applications [41]. Additionally, compatibility with existing manufacturing processes can be an obstacle. In construction, switching from Ordinary Portland Cement (OPC) to geopolymer cement requires alterations in mixing and curing procedures, which can slow down adoption despite its environmental benefits [42]. Furthermore, scalability and consistency in material quality pose significant barriers. Bio-based materials like bamboo and hemp often show variability depending on environmental factors, while recycled metals may contain impurities that affect their mechanical properties, making them less suitable for precision-demanding industries such as aerospace [43, 44]. Overcoming these challenges will require continued innovation and investment in advanced manufacturing technologies to fully realize the potential of green engineering materials.

References

1. Callister, W. D., Jr. (2018). *Materials science and engineering: An introduction* (10th ed.). Wiley.
2. Ashby, M. (2010). *Materials selection in mechanical design*. Butterworth-Heinemann.
3. Ridho, M. R., Agustiany, E. A., Rahmi, M., et al. (2022). Lignin as green filler in polymer composites: Development methods, characteristics, and potential applications. *Advances in Materials Science and Engineering, 2022,* Article ID 1363481, 33 pages. https://doi.org/10.1155/2022/1363481
4. Van De Velde, K., & Kiekens, P. (2001). Biopolymers: Overview of several properties and consequences on their applications. *Polymer Testing, 99,* 483.
5. Ansari, K. R., & Shalwan, A. (2018). Nanotechnology in corrosion protection: A review. *Nanotechnology Reviews, 7*(4), 345–365. https://doi.org/10.1515/ntrev-2018-0035
6. Smith, J., Johnson, A., & Lee, H. (2022). Enhancing corrosion inhibition using nanostructures. *Journal of Materials Chemistry, 10*(5), 123–135.
7. Sirelkhatim, A., Mahmud, S., Seeni, A., Kaus, N. H. M., Ann, L. C., Bakhori, S. K. M., & Mohamad, D. (2015). Review on zinc oxide nanoparticles: Antibacterial activity and toxicity mechanism. *Nano-Micro Letters, 7*(3), 219–242.
8. El-Sheikhy, R., & Al-Shamrani, M. (2017). Interfacial bond assessment of clay-polyolefin nanocomposites CPNC on view of mechanical and fracture properties. *Advances in Powder Technology, 28,* 983–992.
9. Assaedi, H., Shaikh, F. U. A., & Low, I. M. (2016). Effect of nano-clay on mechanical and thermal properties of geopolymer. *Journal of Asian Ceramic Societies, 190,* 19–28.
10. Du, K., Li, P., & Yan, Z. (2019). Do green technology innovations contribute to carbon dioxide emission reduction? Empirical evidence from patent data. *Technological Forecasting and Social Change, 146,* 297–303.
11. Aabid, A., & Baig, M. (2023). Sustainable materials for engineering applications. *Materials, 16*(18), 6085. https://doi.org/10.3390/ma16186085
12. Fontana, M. G. (1986). *Corrosion engineering* (3rd ed.). McGraw-Hill.
13. Uhlig, H. H., & Revie, R. W. (1985). *Corrosion and corrosion control: An introduction to corrosion science and engineering* (3rd ed.). Wiley.
14. Wang, Q., Zhou, J. P., Wang, S., Yang, G. L., & Li, X. (2019). Hydrophobic self-healing polymer coatings from carboxylic acid- and fluorine-containing polymer nanocontainers. *Colloids and Surfaces A: Physicochemical and Engineering Aspects, 569,* 52–58. https://doi.org/10.1016/j.colsurfa.2019.02.050
15. Shchukina, E. H., Wang, W., & Shchukin, D. G. (2019). Nanocontainer-based self-healing coatings: Current progress and future perspectives. *Chemical Communications, 55,* 3859–3867. https://doi.org/10.1039/c8cc09982k
16. Kahia, M., et al. (2016). Impact of renewable and non-renewable energy consumption on economic growth: New evidence from the MENA Net Oil Exporting Countries (NOECs). *Energy,* 116, 102–115.
17. Li, Z., et al. (2019). Structural transformation of manufacturing, natural resource dependence, and carbon emissions reduction: Evidence of a threshold effect from China. *Journal of Cleaner Production,* 206, 920–927

18. Wang, M., et al. (2017). Analysis of energy-related CO2 emissions in China's mining industry: Evidence and policy implications. *Resources Policy*, 53, 77–87.
19. Rice husk. Retrieved from https://scholar.google.com/scholar?hl=en&as_sdt=0%2C5&q=RICE+HUSK+&oq=rice+
20. Palm oil fibre. Retrieved from https://scholar.google.com/scholar?hl=en&as_sdt=0%2C5&q=PALM+OIL+FIBRE&btnG=
21. Bamboo. Retrieved from https://scholar.google.com/scholar?hl=en&as_sdt=0%2C5&q=BAMBOO&oq=ba
22. Reeder, L. (2010). *Guide to green building rating systems: Understanding LEED, green globes, energy star, the national green building standard, and more.* John Wiley & Sons, Inc.
23. Fowler, K. M., & Rauch, E. M. (2006). *Sustainable building rating systems summary.* Pacific Northwest National Laboratory.
24. Roderick, Y., et al. (n.d.). *A comparative study of building energy performance assessment between LEED, BREEAM, and green star schemes.* Integrated Environmental Solutions Limited.
25. Muhit, I. B., Omairey, E. L., & Pashakolaie, V. G. (2024). A holistic sustainability overview of hemp as building and highway construction materials. *Building and Environment, 256*, 111470.
26. True cost of the cement, steel, and chemical industries. Final report, July 2021.
27. Statista. (2021). *Cement prices in the United States from 2007 to 2020.* https://www.statista.com/statistics/219339/us-prices-of-cement/
28. McLellan, B. C., Williams, R. P., Lay, J., van Riessen, A., & Corder, G. D. (2011). Costs and carbon emissions for geopolymer pastes in comparison to ordinary Portland cement. *Journal of Cleaner Production, 19*(9–10), 1080–1090.
29. World steel prices (2020). *European steel prices.* https://worldsteelprices.com/europeansteel-prices/
30. European Commission. (n.d.). *Automotive industry.* https://ec.europa.eu/growth/sectors/automotive_en
31. Mohammadhosseini, H., Alyousef, R., & Md. Tahir, M. (2021). Towards sustainable concrete composites through waste valorization of plastic food trays as low-cost fibrous materials. *Sustainability, 13*(4), 2073. https://doi.org/10.3390/su13042073
32. Awoyera, P. O., Akinmusuru, J. O., & Ndambuki, J. M. (2016). Green concrete production with ceramic wastes and laterite. *Construction and Building Materials, 117*, 29–36.
33. Mytafides, C. K., & Tzounis, L. (2021). Fully printed and flexible carbon nanotube-based thermoelectric generator capable for high-temperature applications. *Journal of Power Sources, 507*, 230323.
34. Zhang, T., Shen, J., Lü, L. Q., Wang, C. M., Sang, J. X., & Wu, D. (2017). *Effects of graphene nanoplates on microstructures and mechanical properties of NSA-TIG welded AZ31 magnesium alloy joints.* Transactions of Nonferrous Metals Society of China (English Ed.). https://doi.org/10.1016/S1003-6326(17)60149-3
35. Alzebdeh, K., Hinai, N. A., Safy, M. A., & Nassar, M. (2023). Recycled polymer bio-based composites: A review of compatibility and performance issues. In *Recycled polymer blends and composites: Processing, properties, and applications* (pp. 363–387). Springer International Publishing. https://doi.org/10.1007/978-3-031-37046-5_18

36. Zahan, A. E., Katsarou, S., & Julien, A. (2018). *Environmental benefits of using cross-laminated timber with hempcrete insulation in buildings*. International Conference for Sustainable Design of the Built Environment (SDBE 2018): Proceedings.
37. Han, D., & Hu, C. (2024). Clinching of carbon fiber-reinforced composite and aluminum alloy. *Metals, 14*(6), 681. https://doi.org/10.3390/met14060681
38. Seow, Y., et al. (2016). A 'design for energy minimization' approach to reduce energy consumption during the manufacturing phase. *Energy*, 109, 894–905.
39. Du, L. M., et al. (2012). Economic development and carbon dioxide emissions in China: Provincial panel data analysis. *China Economic Review*, 23, 371–384
40. Li, K., et al. (2015). Impacts of urbanization and industrialization on energy consumption/CO2 emissions: Does the level of development matter? *Renewable and Sustainable Energy Reviews*, 52, 1107–1122
41. Miele, M., Smith, R., & Johnson, A. (2017). Bio-based composites: Applications in extreme environments. *Journal of Material Science, 45*(6), 1234–1248.
42. Davidovits, J. (2014). *Geopolymer chemistry and applications*. Institute for Research on Geopolymer Science.
43. Ochoa, R., Martínez, A., & López, D. (2017). Recycled metals in aerospace manufacturing: A review. *Aerospace Materials Journal, 21*(3), 456–462.
44. Sandelowski, M., Leavy, P., & Jones, C. (2019). The impact of material variability in green engineering applications. *Journal of Sustainable Materials, 34*(1), 80–91.

2

Bamboo: A Future Renewable Materials Engineering Resource

2.1 Introduction

The construction industry consumes 40–50% of global energy, with a high demand for fossil fuels driven by economic growth in various countries [1, 2]. This increase in construction activities leads to significant CO_2 emissions and other pollutants. Considering sustainable and renewable resources, ensuring a sustainable future requires a pressing need for renewable construction materials. One promising option is bamboo, which meets these criteria by offering a sustainable alternative due to its rapid growth (matured in 3–5 years), high strength-to-weight ratio, and minimal environmental impact [3]. Bamboo, as a cost-effective material, is also highly effective at absorbing and storing CO_2 from the atmosphere. Its energy consumption per unit of production and carbon emissions are only 0.25 kg CO_2 per kilogramme, which is significantly lower than that of steel, which ranges from 2.2 to 2.8 kg CO_2 per kilogramme [4]. Bamboo, recognized since ancient times for its versatility, finds one of its primary applications in the construction industry. It is available in abundance due to its rapid growth and widespread cultivation worldwide, thriving across various altitudes and location [5]. Bamboo is found in tropical and subtropical regions throughout *South America, Asia, Africa, Australia, and the southern parts of North America*. There are approximately 1,200 species of bamboo, with 65 of these species suitable for construction and scaffolding applications. In *Latin America, Guadua angustifolia* is the most renowned species. In Asian countries, including India, popular bamboo species for construction include *Bambusa nutans, Dendrocalamus strictus, Dendrocalamus hamiltonii, Dendrocalamus asper, Bambusa balcooa, Bambusa vulgaris, and Phyllostachys bambusoides* [6]. Other key reasons for its usage include sustainability, flexibility, lightness, aesthetic appeal, low carbon footprint, and local availability.

Bamboo has been widely applied and used for various purposes, ranging from decorative materials to engineering components. Traditionally, bamboo has been utilized for exterior and interior walls, wall panels, window frames,

DOI: 10.1201/9781003617594-2

ceilings, roofs, and scaffolding. In the engineering field, bamboo serves as an alternative to steel reinforcement in concrete construction for floor walls, harbour foundations, and bridges. Bamboo has proven to be effective in both land and sea environments, including humid conditions. Currently, bamboo is being increasingly applied in various sectors such as the automotive, fabrication, and plumbing industries. Its durability and strength have demonstrated its potential as a composite material. With advances in technology and fabrication, bamboo can be enhanced with additional materials, making it more widely accepted by the broader community.

Although bamboo has many advantages, it also has some limitations. In certain conditions, bamboo is not practical to use due to its anisotropic properties and mechanical inconsistencies [7]. Bamboo is much weaker in the transverse direction compared to the longitudinal direction, making round bamboo stems unsuitable for multi-storey buildings. This necessitates the development of engineered bamboo products (EBP) that are more efficient, durable, and have added value suitable for meeting increasing market demand [8]. Bamboo is also vulnerable to wood rot, breaking, and cracking due to shrinkage, and has relatively low strength. Dimensional changes that result in product damage can occur even in bamboo that has been dried before processing. Additionally, bamboo is susceptible to climate changes, temperature fluctuations, and humidity, requiring special attention during transportation, especially between countries. To meet engineering criteria, knowledge and experience in processing bamboo as a building material becomes crucial. Many methods can be employed to enhance the bonding strength and durability of bamboo against weather changes, including the application of waterproof coatings, pest eradication, profile shaping, drying, selection of bamboo types, harvesting methods and timing, bamboo segment selection, growth cycles, proper cultivation methods, and integration with other materials.

The mechanical properties of bamboo, including compression, tension, and modulus of elasticity, are compared among different materials such as steel pipes, wooden rods, and C50 concrete in the **Table 2.1**. These properties demonstrate that bamboo has a superior weight-to-strength ratio compared to other materials.

The woody bamboos are classified into three major groups based on their geographical and altitudinal distribution: paleotropical woody bamboos, neotropical woody bamboos, and north temperate woody bamboos (see **Figure 2.1**). The paleotropical woody bamboos, which include subtribes such as *Bambusinae, Hickelinae, Melocanninae, Racemobambosinae,* and a few other unplaced genera, are native to the Western Hemisphere. They are found in tropical and subtropical regions of Africa, Madagascar, India, Sri Lanka, Southeast Asia, southern China, southern Japan, and Oceania.

TABLE 2.1

Geometry and Strength of Bamboo/Steel Pipe/Wooden Rod [9]

Bar	Externl Diameter D (mm)	Wall Thickness t (mm)	Materil Density y ρ (kg/m³)	Mod. of Elasticity E(GPa)	Ultimate Compression stress σ_{cr},c (MPa)	Ultimae Shear Stress τcr (MPa)
Bamboo bar	90	7	700	14.3	70	7
Steel pipe	60.3	1.65	7845	210.0	245	135
Wooden rod (C50)	70	–	950	22	70	13

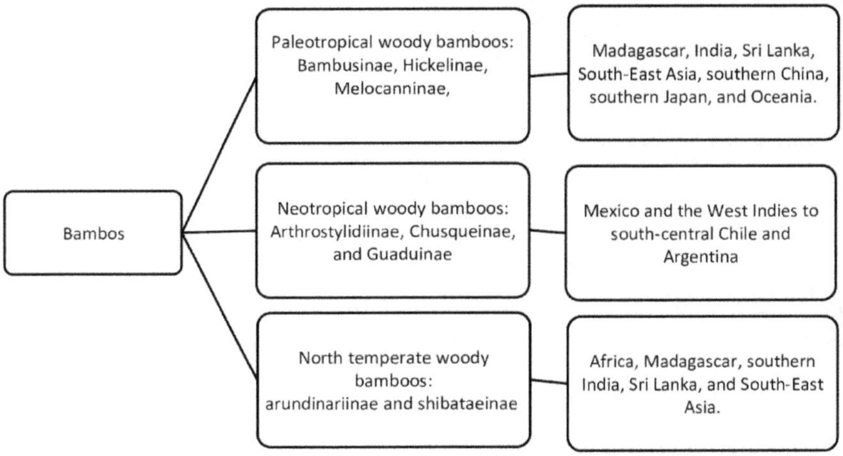

FIGURE 2.1
Classification of bamboo species based on location [10].

The type of bamboo that is widely used in construction are [11]:

- *Timber Bamboo*: this bamboo has large, strong stems and is often used as a construction material. Big bamboo has physical characteristics that make it suitable for various construction projects, such as roofing, walls, floors, and other building structures. Some examples of big bamboo commonly used in construction include:
- *Bambu betung* (*Gigantochloa apus*): this type of bamboo grows in various tropical and subtropical regions and is often used in construction in Southeast Asia. *Betung bamboo (Dendrocalamus asper Backer)* is categorized as a large bamboo with a lower stem diameter of 26 cm and a height of 25 m. Its strong and tough stems make it an ideal material for construction and building materials [12].

- *Bambu petung (Dendrocalamus asper)*: this bamboo is found in various Southeast Asian countries and has large and sturdy stems.
- *Bambu tali (Gigantochloa atroviolacea)*: this bamboo is frequently used in construction due to its large and durable stems.

2.2 Characteristics of Bamboo

2.2.1 Anatomy of Bamboo

The anatomy of bamboo is composed of various parts, each with unique properties. The outer layer, or skin, enhances strength, while the inner layer influences the overall quality of the bamboo. Additionally, the strength of the bamboo varies between the top and bottom sections, impacting its capacity to endure different loads and pressures. To maximize the use of bamboo, it is crucial to consider each part and its specific characteristics. Thus, understanding bamboo anatomy is vital for selecting the appropriate components for particular applications and environmental conditions.

Bamboo is a natural composite made up of cellulose fibres embedded within a reinforced bamboo plant. These cellulose fibres significantly enhance the bending and tensile strength of bamboo, with tensile strength being highest along the direction of the fibres [13]. The fibres form the vascular tissue within the stem, essential for transporting nutrients to bamboo shoots, and are bonded together by lignin. Both cellulose and lignin fibres play crucial roles in augmenting bamboo's strength, making it comparable to other natural composites [14].

Bamboo can be utilized in various ways depending on its parts (see **Figure 2.2**). The leaves are used for manure, fodder, and extracting medicine. Twigs are crafted into brooms and cloths. The top sections of bamboo are ideal for making chopsticks, toothpicks, and scaffolding. The middle upper parts are used for mats, carpets, chopsticks, and handicrafts, while the middle lower sections are suitable for flooring and laminated furniture. Finally, the base of the bamboo can be processed into charcoal and pulp [15].

2.3 Mechanical Aspects of Bamboo

The mechanical properties of bamboo are influenced by several factors. These factors include the type of bamboo, direction, and age. These factors also affect the density of the bamboo. Fibres at specific points within the bamboo determine its strength. Bamboo has different mechanical

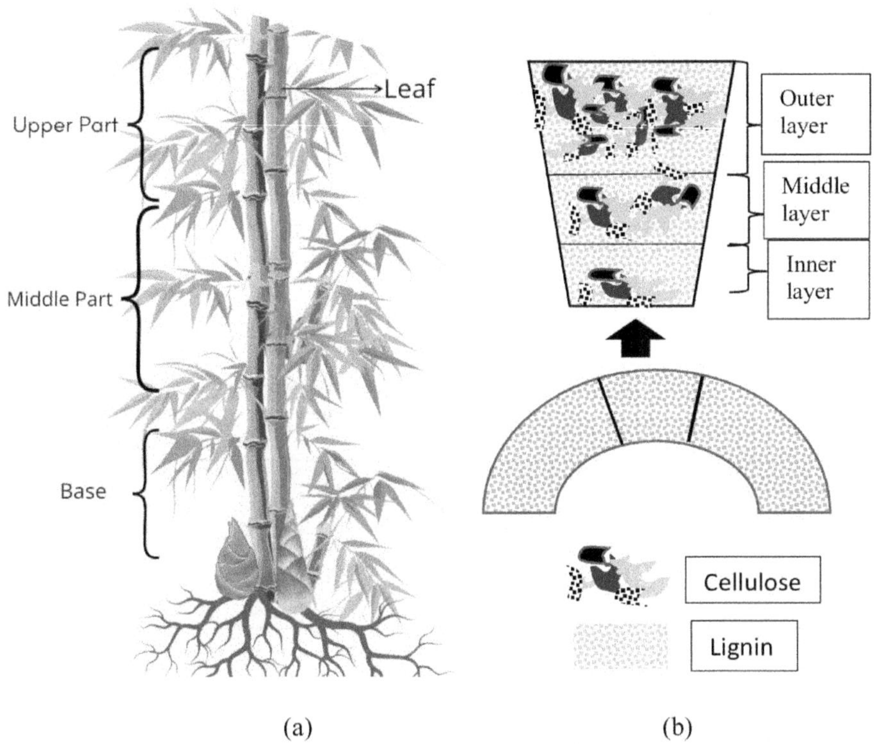

FIGURE 2.2
Anatomy of bamboo: (a) part of the bamboo, (b) cross-section of bamboo [15] (Illustration: canva, 2024, https://www.canva.com).

characteristics along its longitudinal, radial, and tangential directions (Sri Handayani). Considering its applications in various engineering and construction contexts, there are critical factors that significantly influence the overall strength of bamboo, including the moisture content, the configuration of the transverse section, and the presence or absence of nodes within the bamboo structure [16]. In general, the mechanical properties of bamboo are presented in **Table 2.2**.

When comparing the mechanical properties of bamboo with other materials such as mild steel, polymer resin, GRP with WR and CSM, and unidirectional GRP, several key differences and similarities can be observed in the **Table 2.3** below.

The strength of bamboo is influenced by several factors, including its age (**Table 2.4**), the specific part of the bamboo being used, and the internodal modulus. Younger bamboo tends to be less strong than older bamboo, which has had more time to develop denser, more resilient fibres. Different parts of the bamboo also exhibit varying strengths; for example, the outer layer is typically stronger than the inner layers, and the lower sections of the bamboo

TABLE 2.2

Mechanical Properties of Bamboo [16]

Mechanical Properties	Strength (Kg/cm^2)
Tensile	981–3,920
Flexture	686–2,940
Compressive	245–981
E (Modulus)	98,070–29,4200

TABLE 2.3

Tensile and Compression Test Results [17]

Material	Specific Modulus (km^2/s^2)	Specific Tensile Strength (km^2/s^2)	Specific Compressive Strength (km^2/s^2)
Mild steel	25,316	50.6	–
Polymer resin	3,636	36.4	90.9
GRP with WR and CSM	4,965	103.4	89.1
Unidirectional GRP	22,944	250	166.7
Bamboo	22,889	214.4	75.96

TABLE 2.4

Compression Test Results [19]

Year	High	SG	Longitudinal	
			fc (MPa)	*Ec* (MPa)
One	Bottom	0.49	47.0 (2.4)	2067 (339)
	Middle	0.53	50.9 (3.1)	2776 (362)
	Top	0.54	55.7 (3.8)	3658 (464)
Three	Bottom	0.70	86.8 (1.8)	4426 (491)
	Middle	0.71	83.9 (2.8)	4428 (305)
	Top	0.72	84.0 (3.3)	4660 (451)
Five	Bottom	0.75	93.6 (3.6)	4896 (116)
	Middle	0.78	86.6 (3.5)	4980 (262)
	Top	0.76	85.8 (5.3)	5185 (330)

are generally stronger than the upper sections. Additionally, the internodal modulus (see **Figure 2.3**), which refers to the compressive stress and elasticity of the sections between the nodes, plays a crucial role in determining the overall strength of the bamboo. Understanding these factors is essential for optimizing the use of bamboo in construction and other applications, ensuring that the right type and section of bamboo are chosen for specific structural needs. The concentration of cellulose fibres rises gradually from the interior to the exterior of the culm. This non-uniform arrangement of cellulose fibres will have an impact on the mechanical characteristics of the

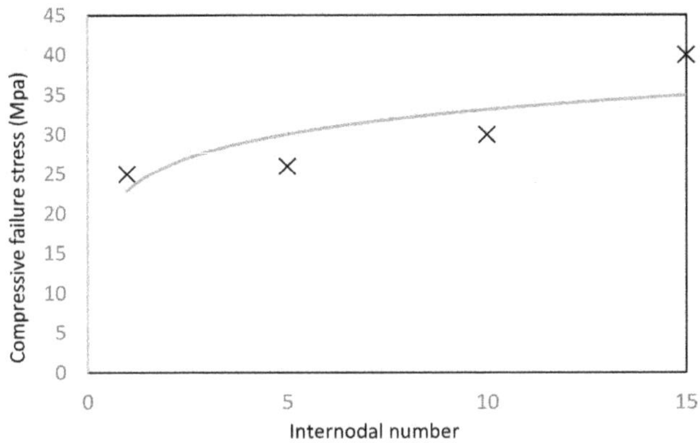

FIGURE 2.3
Plot of ultimate compressive strength versus distance of bamboo culm from ground [20].

bamboo. Typically, the cellulose fibres make up 40% of the culm's volume overall [18].

2.3.1 Density of Bamboo

Bamboo is classified as a hygroscopic material, capable of absorbing water in both vapour and liquid forms. Wood and bamboo have the ability to absorb or release moisture, depending on temperature and humidity. The moisture content within the bamboo stem varies both longitudinally and transversely. This variability is influenced by factors such as the age, time of harvesting, and type of bamboo. At one year of age, the bamboo stem typically exhibits a relatively high moisture content, ranging from approximately 120% to 130% at both the base and the top [21].

Density of wood or bamboo is a determining factor for their physical and mechanical properties. The specific gravity of bamboo ranges between 0.5 and 0.9 gr/cm^3 [22]. The density of bamboo varies both vertically and horizontally. **Figure 2.4** shows the variation in bamboo fibre density based on the distance from the inner surface. The outer part of the bamboo exhibits a higher density in comparison to the inner part. Longitudinally, the density tends to increase from the base to the top. Density is inversely correlated with moisture content, meaning that as the density of bamboo increases, its moisture content decreases [23].

2.3.2 Bamboo Treatment

Naturally, bamboo has low durability and is susceptible to attack by destructive organisms such as dry wood powder beetles, drywood termites, and subterranean termites. Bamboo's anatomical properties are similar to those of wood, so factors affecting wood also influence bamboo. The durability of

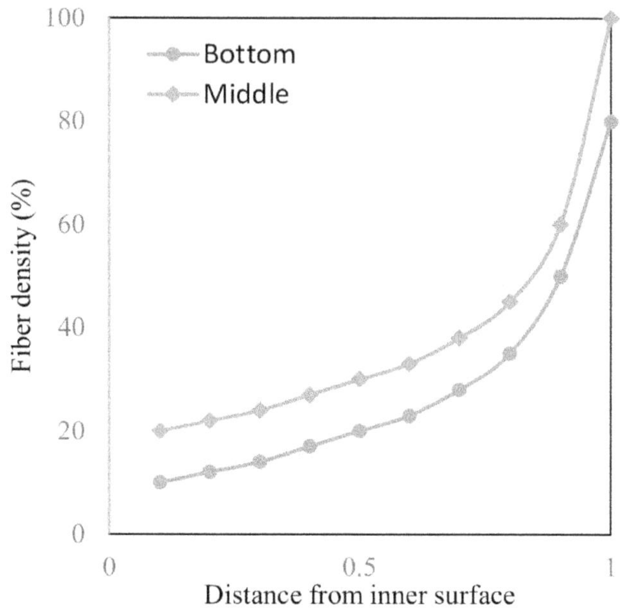

FIGURE 2.4
Plot of fibre density percentage versus distance from the inner surface of bamboo [24].

bamboo is influenced by its water content and specific gravity [16]. Bamboo is a hygroscopic material, meaning it can absorb moisture from the air in the form of vapour or liquid, depending on the temperature and humidity. Thus, to ensure durability and consistent strength, bamboo requires specific treatments to address these destructive factors. One of the principles of bamboo preservation involves removing the starch content that allows powder beetles to thrive and grow, or reducing bamboo porosity through chemical or thermal methods. These treatments can include chemical preservatives, thermal treatments, and physical modifications. Bamboo preservation methods can be divided into two categories: chemical and non-chemical [16].

Non-chemical methods include soaking bamboo in water to prevent powder beetle attacks. Another non-chemical method involves boiling the bamboo at a temperature of 100°C for one hour, followed by pressure treatment to ensure water is evenly distributed across the segments. During this process, the pressure is carefully regulated to avoid breaking the bamboo, and it is pumped into a preservation tank.

Chemical preservation involves using substances like diesel oil or borax. In the diesel oil method, freshly cut bamboo is placed upside down with its tip attached to a tube filled with diesel oil. The gravity-driven flow of diesel oil forces the liquid out of the bamboo. Alternatively, borax (*natrium tetraborat*/$Na_2B_4O_7$) can be used in place of diesel oil for preservation. **Table 2.5** and **Table 2.6** compare the moisture content and compression strength in Ori bamboo and Wulung bamboo at different sections (base, middle, and tip)

TABLE 2.5

Comparison Moisture Content before and after Borax Treatment [16]

	Moisture Content Ori (%)		Moisture Content Bambu Wulung (%)	
	No Borax	**With Borax**	**No Borax**	**With Borax**
Base	36.357	26.32	63.267	39.78
Middle	33.629	24.00	50.805	25.81
Tip	32.422	23.67	42.539	23.26

TABLE 2.6

Comparison Compressive Strength before and after Borax Treatment [16]

	Comp. Strength Bambu Ori (kg/cm^2)		Comp. Strength Bambu Wulung (kg/cm^2)	
	No Borax	**With Borax**	**No Borax**	**With Borax**
Base	3,060.67	2,553.95	4,473.74	1,255.14
Middle	3,088.38	2,613.79	4,133.61	1,477.66
Tip	3,233.61	2,796.43	3,191.69	1,528.19

under two conditions: without borax treatment and with borax treatment. Borax treatment is effective in reducing the moisture content and increasing the compression strength of both Ori and Wulung bamboo. This reduction in moisture can help enhance the durability and resistance of bamboo to pests and decay, making it more suitable for various applications. The data highlights the significant impact of borax on lowering the moisture levels, particularly in *Wulung bamboo,* which initially has higher moisture content compared to *Ori bamboo.*

Traditionally, bamboo preservation methods have utilized natural resources and techniques such as water, smoke, and other organic materials. Water-based preservation involves soaking the bamboo in water to leach out starches and sugars that attract pests. This method can take several weeks but is effective in reducing the bamboo's susceptibility to insect attacks. The following are examples of traditional methods used to preserve bamboo [25]:

2.3.2.1 Water-Leaching

Soaking is one of the simplest techniques for preserving bamboo. This technique involves submerging bamboo in clean flowing water for approximately 4–8 months. Through this process, substances present in bamboo, such as starch and sugar, dissolve along with the water. This reduces the likelihood of insect attacks, as these substances serve as food for insects. However, bamboo is still not completely resistant to insect and mould attacks with this soaking process.

2.3.2.2 Fermentation

The fermentation technique is carried out by placing treated bamboo in mud and tree leaves for 3–4 months. Through this process, microorganisms and bacteria in the compost are able to convert the starch and sugar content from the bamboo into acid. This acidic environment is less attractive to insect attacks. However, this method is less consistent and does not offer complete protection for bamboo from long-term insect infestations.

2.3.2.3 Smoking and Heating

Smoking bamboo is essential to reduce the moisture content in newly harvested bamboo, which typically retains high levels of moisture and sugar content. During the smoking process, chemical compounds present in the smoke react with the bamboo tissue, providing protection against insect attacks. However, the challenge lies in maintaining consistent temperature and smoke quality throughout the preservation process to ensure uniform treatment for all parts of the bamboo. Implementing this process on a large scale is complex, particularly due to its environmental impact. Heat treatment is also can be employed to remove moisture that could lead to decay. The drying process must be carried out carefully to maintain the bamboo's integrity and prevent excessive shrinkage.

2.3.2.4 Salt Water/Sea Water Immersion

Soaking bamboo in saltwater for a duration of 30 days to three months enables the absorption of salt into the bamboo fibres, facilitating the release of the sweet sap contained within the bamboo. Creating perforations in the core of the bamboo facilitates the optimal penetration of salt into the bamboo, particularly for bamboo with thick walls. An alternate method involves drilling small holes from the outer skin at each internode. However, the main challenge is corrosion, as seawater comprises sodium and chlorine ($NaCl$), which can deteriorate construction materials, particularly steel used for construction joints, ultimately leading to rusting. This corrosion process is further accelerated by the presence of water absorbed from the surrounding air, potentially causing long-term structural issues.

2.4 Chemically Treated Bamboo

2.4.1 Urea

Research utilizing urea at concentrations of 20%, 40%, and 60% was conducted to evaluate its impact on dimensional stabilization [26]. The effectiveness of this stabilization is measured through various parameters: shrinkage in both tangential and radial directions, anti-shrink efficiency (ASE) values for tangential and radial directions, and the comparison between tangential and radial shrinkage (T/R ratio). The results can be explained as follows. Bamboo apus, when stabilized with a 20% concentration of urea, exhibited reduced tangential and radial shrinkage at 0.684% and 1.676%, respectively, in comparison to the non-stabilized sample. The anti-shrink efficiency (ASE) in the tangential direction reached 83.766%, while in the radial direction, it was 69.335%. On the other hand, bamboo petung required a 60% urea concentration to achieve decreased tangential and radial shrinkage, measuring at 0.372% and 1.383%, respectively. The ASE in the tangential direction was 71.213% and in the radial direction, it was 86.589%.

2.4.2 Other Chemicals

Copper chromium arsenate (CCA), copper chromium boron (CCB), permethrin, and pyrethroids are examples of chemicals that can be used to treat bamboo, significantly improving its durability in various atmospheric conditions. Although they are highly effective in killing microorganisms and fungi, the use and disposal of these chemicals pose significant environmental risks. It is crucial to implement waste management and regulatory practices before employing this technique [27].

2.4.3 Hygrothermal Treatment

Hygrothermal treatment refers to the process of treating materials using a combination of heat (thermal) and moisture (hygro). This method is often employed to enhance the physical and mechanical properties of materials, particularly wood and bamboo. The process typically involves exposing the material to high temperatures and controlled humidity levels for a specified period. Hygrothermal treatment has been applied for a long time and continues to develop because it is environmentally friendly and efficient. In research on *Moso bamboo* using the hygrothermal method, at a temperature of 180°C and 100% relative humidity (RH), the treatment produced the highest crystallinity (36.92%), an increase of 11.07% compared to the control sample [28]. The treatment also resulted in a reduction in pore size and average pore diameter (2.72 nm), decreasing by approximately 38.2%. Additionally, there was an increase in the hardness of the samples by almost 21% compared to the control samples. It can be concluded that 180°C is a suitable hygrothermal treatment temperature for *Moso bamboo* modification due to the changes in porosity and increased cell wall nanomechanics.

When selecting a treatment type or chemical and application method for bamboo, several important considerations must be taken into account [29]. The quantity of bamboo to be treated and the availability of treatment facilities are crucial factors. Additionally, the availability of chemicals, the

TABLE 2.7

Recommended Time for Bamboo Treatment in Warm and Aggressive Environments with High Risk of Termite and Beetle Infestation [31]

	Untreated	Treated with Boron	Treatment with Fixed Preservatives
Internal	2–6 years	30+ years	30+ years
External above ground	0.5–4 years	2–15 years	30+ years
External in-ground contact	< 0.5 years	< 1 year	15+ years

intended use of the bamboo (whether inside or outside), and adherence to country legislation must be considered. The species of bamboo is also important, as some species are more readily treatable than others. Transport time from the harvest location to the treatment facility is another critical factor, as some treatment methods require very freshly cut bamboo. Budget constraints and the effectiveness of the treatment type or chemical and application method should be evaluated. Furthermore, it is important to consider whether the chemical affects the structure of the bamboo or any metal fastenings, as well as the toxicity of the chemical to humans throughout its whole life cycle (treatment, use, and disposal) [30]. **Table 2.7** provides guidelines on the suggested durations for bamboo treatment based on various environmental conditions and the type of treatment used.

2.4.4 Bamboo Photostability

Photostability refers to the ability of bamboo to maintain its surface properties when exposed to light, especially ultraviolet (UV) light. This is crucial for ensuring the quality of bamboo used in decorative applications such as flooring, panels, and furniture, by preventing significant physical and chemical changes. Key strategies to achieve photostability can be described as follows:

1. Organic UV absorbers (acrylic coatings): incorporating 3 wt% *benzophenone* (*BP*) *and benzotriazole* (*BTZ*) in the acrylic coatings has been found to effectively protect bamboo from UV-accelerated ageing. These coatings can inhibit lignin photodegradation and hydroxyl group formation, providing protection for 380 and 150 hours, respectively [32].
2. Inorganic UV absorbers (metal oxide nanoparticles): nanoparticles such as *zinc oxide* (ZnO) are commonly used for their ability to absorb or scatter light without undergoing photodegradation. With particle sizes ranging from 20 nm to 50 nm, these nanoparticles can absorb UV light and visible light at wavelengths of 400–500 nm.

ZnO/polypropylene nanocomposite coatings have been noted to significantly reduce the rate of photodegradation while enhancing the ductility and tensile strength of the coatings [33].

3. Hindered amine light stabilizers (HALS) (combination with organic UV absorbers): HALS are often combined with organic UV absorbers and antioxidants to enhance photostability. They have been proven to protect phenol-formaldehyde (PF) resins from photodegradation and improve the quality of the surface layer on wood. The combination of organic UV absorbers and HALS additives preserves the natural colour of the wood by converting UV energy into heat before it reaches the substrate [34].

2.4.5 Engineered Bamboo

- Bamboo-reinforced concrete.

Bamboo has the potential to be used as a reinforcing element to replace steel in reinforced concrete structures, resulting in an approximate 4% increase in the concrete's strength. To enhance the adhesion between the bamboo and the concrete, the application of water-repellent coatings is recommended. Coating bamboo-reinforced concrete (BRC) with fire retardant materials enables the material to endure extreme temperatures of 500°C or more. These attributes highlight the advantages of BRC panels, particularly in the construction of lightweight and cost-efficient walls [1]. **Table 2.8** shows the ultimate stress and maximum deflection of bamboo-reinforced concrete at 28 days of the average of three samples.

- Laminated bamboo

The laminated bamboo is one of the modifications of bamboo that is beneficial for improving mechanical and physical properties. It can be produced in various forms, standard sizes, and with improved characteristics, making it versatile for structural applications. There are several factors that affect the quality of laminated bamboo, including the type of adhesive, surface roughness, adhesive quantity, pressure load, and the type of joint. The use

TABLE 2.8

Maximum Ultimate Load and Deflection of Concrete and Modified Concrete [35]

Sample	Ultimate Load (kN)	Maximum Deflection (mm)
Concrete	12.5 0	0.26
Singly reinforced beam	22.4	1.18
Doubly reinforced beam	30.0	2.12

of phenol-formaldehyde (PF) adhesive on both sides with a quantity of 250 gm^{-2} resulted in the best mechanical properties for structural purposes [36].

2.5 Bamboo Biocomposites

Bamboo fibre-reinforced biocomposites offer distinct properties that surpass those of standalone bamboo, such as improved flexural tensile strength, flexibility, resistance to cracking, as well as impact strength and toughness. These characteristics are highly significant as they enable broader applications in various fields. Examples and classifications of bamboo biocomposites are mentioned below [37].

2.5.1 Conventional Biocomposites

- Chipboard and flakeboard.
- Plywood and laminated.
- Medium density fibreboard.
- Medium density fibreboard.

2.5.2 Advanced Polymer Biocomposite

- Thermoplastic-based bamboo composites.
- Thermoset-based bamboo biocomposites.
- Elastomer-based biocomposites.

2.5.3 Inorganic-Based Biocomposites

- Gypsum-bonded particleboards.
- Inorganic cement board.

2.6 Bamboo Deterioration

Deterioration in bamboo can result from both biological and non-biological factors. Among the significant non-biological factors is the moisture content. Elevated moisture content can lead to reduced bamboo strength and promote decay. Biologically, bamboo can be damaged by termites, powder post beetles, and fungi, including *Schizophyllum commune, Auricularia sp, Pleurotus sp, Strureum sp, and Poria incrassata sp.* Powder post beetles reside within the bamboo fibre tissue, deriving their nourishment from the starch. The speed of destruction of bamboo by biological activity occurs fastest on fresh green bamboo, which is highly susceptible to damage. Bamboo is attacked not only in wet conditions but even dry bamboo can be attacked, especially in warm and humid climates. In such conditions, most of the water content is in the outer layer of the bamboo, which can then penetrate all parts of the bamboo, making it more susceptible to damage [30].

2.6.1 Fungi

High carbohydrate content in bamboo makes it susceptible to insects and fungal infestations. Fungi that colonize bamboo contribute to its decay by breaking down organic compounds, leading to a deterioration of its structural integrity. This process can result in reduced mechanical strength, alterations in colour, and ultimately a decrease in durability. Different types of fungi can have varying impacts on the lifespan of bamboo. Expanding bamboo's applications requires the development of effective preservation technologies. Experiments involving 16 genera and 18 species of fungi concluded that *Trametes versicolor* and *Arthrinium arundinis* were the most destructive, causing significant weight loss in the test samples, approximately 21.6% and 17.9%, respectively [38]. The study indicates that bamboo's durability is severely compromised when exposed to outdoor air, leading to serious damage from these fungal species. Another report [39] noted that fungal attacks on bamboo (*Gigantochloa scortechinii*) lead to accelerated ageing and decay, significantly reducing its lifespan. The total weight loss and percentage of weight loss caused by white rot and brown rot fungi are detailed in **Table 2.9**. The data shows that bamboo is less resistant to brown rot decay compared to white rot. This is because white rot fungi can utilize both carbohydrates and lignin, while brown rot fungi only modify lignin during the decay process.

TABLE 2.9

Bamboo Decay after Eight Weeks of Fungi Exposure [39]

	TWC* (mg)		PWL* (%)	
Age (years)	**White Rot**	**Brown Rot**	**White Rot**	**White Rot**
0.5	140	150	9.90	9.95
3.5	120	130	9.24	9.49
6.5	60	90	5.30	8.90

**TWC:* total weight loss, *PWL: percentage weight loss

2.6.2 Photodegradation

Photodegradation significantly impacts bamboo, causing a noticeable change in its colour as it fades or darkens over time. This process also alters the chemical composition of bamboo, breaking down its organic components and affecting its overall durability. Additionally, photodegradation modifies the microstructure of bamboo, potentially weakening its structural integrity and making it more susceptible to damage. The surface of bamboo is affected by several environmental factors, including solar radiation (comprising UV light, visible light, and infrared), moisture (dew, rain, snow, and humidity), temperature, oxygen, and fungi [40, 41]. The primary cause of damage to the bamboo surface is the light energy from solar radiation, which triggers various photochemical changes. Bamboo is primarily composed of cellulose (approximately 55%), hemicellulose (around 20%), and lignin (about 25%), each of which absorbs UV rays to different extents. Although UV light constitutes only about 6.8% of sunlight, its high energy levels cause significant damage to the chemical bonds in bamboo components, particularly lignin, at wavelengths of 346, 334, and 289 nm. UV radiation can break single C-C, C-O, and C-H bonds, leading to the degradation of bamboo's structural integrity [42].

An experiment using UV light demonstrated that UV exposure causes photodegradation and photodiscolouration on the surface of bamboo [43]. Specifically, UV 313 and high-pressure mercury light induce significant colour changes in bamboo. These changes occur rapidly at first and then stabilize over an extended period. High-pressure mercury lamp light induces discolouration quickly. As a result of exposure to these light sources, light-coloured bamboo changes to a darker colour, and conversely, dark-coloured bamboo lightens. Surface colour changes in bamboo scrimber occur more slowly as the bamboo has undergone heat treatment, resulting in a denser structure. Bamboo extracts contribute to red colour changes, while lignin contributes to yellow colour changes during photodiscolouration. Dark carbonized scrimber undergoes fewer changes compared to light carbonized scrimber.

References

1. Bala, A., & Gupta, S. (2023). Engineered bamboo and bamboo-reinforced concrete elements as sustainable building materials: A review. *Construction and Building Materials*, 394, 132116.
2. Kucukvar, M., & Tatari, O. (2013). Towards a triple bottom-line sustainability assessment of the US construction industry. *International Journal of Life Cycle Assessment*, *18*(5), 958–972.

3. Joseph, P., & Tretsiakova-McNally, S. (2010). Sustainable non-metallic building materials. *Sustainability*, 2(2), 400–427. https://doi.org/10.3390/su2020400
4. Ashby, M. (2010). *Materials selection in mechanical design*. Butterworth-Heinemann.
5. Gupta, A., & Kumar, A. (2008). Potential of bamboo in sustainable development. *Asia Pacific Business Review, 4*(3), 100–107.
6. Vengala, J., Jagadeesh, H. N., & Pandey, C. N. (2008). Development of bamboo structure in India. In Y. Xiao, M. Inoue, & S. K. Paudel (Eds.), *Modern bamboo structures* (pp. 63–76). CRC Press.
7. Sharma, B., Gatóo, A., Bock, M., & Ramage, M. (2015). Engineered bamboo for structural applications. *Construction and Building Materials, 81*, 66–73.
8. Gonzalez, M. G. (2020). *Fire analysis of load-bearing bamboo structures* (PhD thesis). University of Queensland, School of Civil Engineering.
9. Moreira, L. E., Seixas, M., & Ghavami, K. (2019). *Lightness and efficiency of structural bars: Comparison of bamboo, steel and wood bars under compression*. 18th International Conference on Non-Conventional Materials and Technologies, Construction Materials & Technologies for Sustainability, 18th NOCMAT 2019, Nairobi, Kenya.
10. Kerala Forest Research Institute (KFRI). (2024). https://www.bambooinfo.in/species/worldwide-distribution-of-bamboo.asp
11. Larsen, M. (2023). *Different types of bamboo and their uses in construction*. https://bamboou.com/different-types-of-bamboo-and-their-uses-in-construction.Bambusa%20blumeana%2C%20also%20known%20as,is%20native%20to%20Southeast%20Asia
12. Sutiyono, S., & Wardan, M. (2011). *Characteristics of Bambu Petung (Dendrocalamus asper Back.) plantation in lowland, Subang District, West Java*. Seminar Nasional VIII Pendidikan Biologi FKIP UNS.
13. Lakkad, S. C., & Patel, J. M. (1981). Mechanical properties of bamboo, a natural composite. *Fibre Science and Technology, 14*(4), 319–322.
14. Schröder, S. (2015). *Moso - transverse surface of culm wall*. https://www.guadu-abamboo.com/guadua/comparing-mechanical-properties-of-bamboo-guadua-vs-moso
15. Conduah, G. (2011). *Prospect of utilising bamboo wall-panels in lieu of sandcrete block walling for housing construction in Ghana* (MSc thesis). Heriot-Watt University, School of the Built Environment. https://www.canva.com
16. Handayani, S. (2007). Bamboo mechanical properties. *Teknik Sipil & Perencanaan, 9*(1), 43–53.
17. Lakkad, S. C., & Patel, J. M. (1981). Mechanical properties of bamboo, a natural composite. *Fibre Science and Technology, 14*(4), 319–322.
18. Zou, L., Jin, H., Lu, W.-Y., & Li, X. (2009). Nanoscale structural and mechanical characterization of the cell wall of bamboo fibers. *Materials Science and Engineering: C, 29*(4), 1375–1379.
19. Li, X. (2004). *Physical, chemical, and mechanical properties of bamboo and its utilization potential for fiberboard manufacturing*, Louisiana State University and Agricultural and Mechanical College (PhD thesis).
20. Verma, C. S., & Chariar, V. M. (2012). Study of some mechanical properties of bamboo laminae. *International Journal of Metallurgical & Materials Science and Engineering, 2*(2), 20–37.
21. Pathurahman. (1998). *Aplikasi Bambu pada Struktur Gable Frame* (Master thesis). Fakultas Teknik, UGM, Yogyakarta.

22. Samsudin, M. (1997). *Sambungan Bambu dengan Baut dan Pengisi* (Master's thesis). Program Pasca Sarjana, UGM, Yogyakarta.
23. Wulandari, F. T. (2014). Sifat fisika empat jenis bambu lokal di Kabupaten Sumbawa Barat. *Media Bina Ilmiah, 8*(1), 1–5.
24. Nogata, F., & Takahashi, H. (1995). Intelligent functionally graded material: bamboo. *Composites Engineering, 5*(7), 743–751.
25. Farrugia, M., & Goutham, S. (2022). *Traditional methods for treating bamboo.* https://bamboou.com/traditional-methods-for-treating-bamboo/
26. Kusumaningsih, R. (2012). The increasing of bamboo quality using dimensional stabilization. *Jurnal Wana Tropika, 2*(1). https://jurnal.instiperjogja.ac.id/index.php/JWT/article/view/72/70
27. Chemical Bamboo Preservation. (2024). https://www.guaduabamboo.com/blog/chemical-bamboo-preservation
28. Ye, C., Huang, Y., Feng, Q., et al. (2020). Effect of hygrothermal treatment on the porous structure and nanomechanics of Moso bamboo. *Scientific Reports, 10,* 6553. https://doi.org/10.1038/s41598-020-63524-0
29. Liese, W., Gutiérrez, J., & González, G. (2002). Preservation of bamboo for the construction of houses for low income people. In A. Kumar, I. V. R. Rao, C. Sastry (Eds.), *Bamboo for sustainable development* (pp. 481–494).
30. Kaminski, S., Lawrence, A., Trujillo, D. J. A., & King, C. (2016). Structural use of bamboo: Part 2: Durability and preservation. *The Structural Engineer, 94*(10), 38.
31. Lebow, S. (2004). *Alternatives to chromated copper arsenate for residential construction* (Research Paper FPL-RP-618). U.S. Department of Agriculture, Forest Service, Forest Products Laboratory.
32. Rao, F., Chen, Y., Zhao, X., Cai, H., Li, N., & Bao, Y. (2018). Enhancement of bamboo surface photostability by application of clear coatings containing a combination of organic/inorganic UV absorbers. *Progress in Organic Coatings, 124,* 314–320.
33. Zhao, H., & Li, R. K. Y. (2006). A study on the photo-degradation of zinc oxide (ZnO) filled polypropylene nanocomposites. *Polymer, 47*(9), 3207–3217.
34. Evans, P. D., Kraushaar Gibson, S., Cullis, I., Liu, C., & Sèbe, G. (2013). Photostabilization of wood using low molecular weight phenol formaldehyde resin and hindered amine light stabilizer. *Polymer Degradation and Stability, 98*(1), 158–168.
35. Rashid, H., & Hossain, M. A. (2011). Performance evaluation of bamboo reinforced concrete beam. *Journal of Engineering Technology,* 11(4), 142–146
36. Sulastiningsih. (2021). Physical and mechanical properties of glued laminated bamboo lumber. *Journal of Tropical Forest Science, 33*(3), 290–297.
37. Siti, S., Abdul, H. P. S., Wan, W. O., & Jawai, M. (2013). *Bamboo based biocomposites: Material, design, and applications.* InTech.
38. Kim, J.-J., Lee, S.-S., & Lee, A. H. (2011). Fungi associated with bamboo and their decay capabilities. *Holzforschung, 65,* 271–275.
39. Erickson, K. E. (1978). Enzyme mechanism involved in cellulose hydrolysis by the white rot fungus, Sporatichum pulverulentum. *Biotechnology and Bioengineering, 20,* 317–332.
40. Wang, X. Q., & Ren, H. Q. (2009). Surface deterioration of Moso bamboo (Phyllostachys pubescens) induced by exposure to artificial sunlight. *Journal of Wood Science, 55*(1), 47–52.

41. Kim, Y. S., Lee, K. H., & Kim, J. S. (2016). Weathering characteristics of bamboo (Phyllostachys pubescens) exposed to outdoors for one year. *Journal of Wood Science, 62*(4), 332–338.
42. Rao, F., Li, X., Li, N., Li, L., Liu, Q., Wang, J., Zhu, X., & Chen, Y. (2022). Photodegradation and photostability of bamboo: Recent advances. *ACS Omega, 7*(28), 24041–24047.
43. Yu, H., He, S., Zhang, W., Zhan, M., Zhuang, X., Wang, J., & Yu, W. (2021). Discoloration and degradation of bamboo under ultraviolet radiation. *International Journal of Polymer Science*, 2, 1–10

3

Plant-Based Composites: A Sustainable Resource for Future Engineering

3.1 Introduction

The continuous decline in global petroleum reserves, the rise in petroleum-based plastic prices, along with the rapid demand for industrialization, needs a steady supply of engineering materials. This is becoming a crucial situation and must be addressed seriously. To ensure an adequate supply of materials, it is important to boost production and manufacturing activities. As these activities intensify, there is an increase in material and energy usage, resulting in increasing waste generation that poses a significant environmental impact. In response to the worsening environmental issues, the use of biodegradable-based materials needs to be developed [1, 2]. Researchers are actively seeking to implement sustainable materials as alternatives. Presently, they are actively exploring sustainable and environmentally friendly materials that adhere to stringent engineering standards. This comprehensive effort involves the utilization of plant-based and renewable resources, marking a significant step towards eco-conscious composite material development.

Composite materials are composed of two or more elements with complementary properties, where one component acts as an adhesive and the other as reinforcement. These composites possess strengths such as being lightweight, corrosion-resistant, and visually appealing. Biocomposites offer several advantages, including cost-effectiveness, desirability, lightweight attributes, non-abrasive properties, non-toxicity, and notably, these materials have also demonstrated the ability to meet engineering criteria, particularly in terms of their durability and toughness. By utilizing natural resources such as rice husk waste, various types of leaves, stems, and plant roots, impressive results have been shown in terms of tensile strength, flexibility, high-temperature resistance, and resistance to radiation exposure [3, 4]. Consequently, these materials hold promise for potential applications in replacing conventional metals and alloys, thus promoting a more sustainable industrial landscape.

DOI: 10.1201/9781003617594-3

3.2 Composites

Composite is an alloy material consisting of a combination of the main material (matrix) and reinforcement (fibre). The matrix and fibre have complementary mechanical properties. Composites are formed by combining these mechanical properties to obtain combined mechanical properties. Based on the matrix, composites are divided into three types [5, 6], namely:

- Polymer matrix composites (PMC).
- Metal matrix composites (MMC).
- Glass and ceramic matrix composites (GCMC).

Table 3.1 shows a variety of materials used for fibre reinforcement in composites and describes their specific roles in enhancing the mechanical properties within the composite materials.

Figure 3.1-a presents typical composite materials consisting of a combination of matrix materials and reinforcement fibres. **Figure 3.1-b** shows the mechanical properties of composites, showcasing the combination of the mechanical properties of the materials, with fibres contributing to ductility, while the matrix provides hardness and toughness.

In composite materials, the fibre structure within the matrix significantly impacts mechanical properties. These structures include continuous fibres, random fibres, cross-ply laminates, and sandwich structures, each enabling composites to meet diverse engineering needs effectively. **Figure 3.2** visually displays the configuration of typical composite elements, illustrating the application of honeycomb reinforcement (a) as well as a fabricated sandwich panel (b). Honeycomb reinforcement is commonly used as a structural core in aerospace and marine applications to provide exceptional strength and

TABLE 3.1

The Materials Used for Fibre Production Applicable in Industries and Their Corresponding Mechanical Properties [5, 6]

Fibres	Mechanical Properties
Glass	High strength, low stiffness, high density, low cost; E (calcium aluminoborosilicate) and S (magnesia-aluminosilicate).
Graphite	High modulus and strength, low cost, lower density than glass.
Boron	High strength and rigidity, highest density; the highest cost; has a tungsten filament in the middle.
Aramids (Kevlar)	Highest strength-to-weight ratio, high cost.
Etc	Nylon, silicon carbide, silicon nitride, aluminium oxide, boron carbide, boron nitride, tantalum carbide, steel, tungsten, molybdenum.

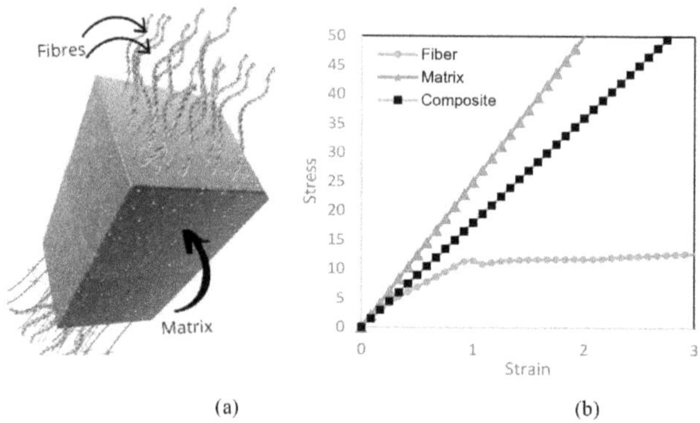

FIGURE 3.1
Illustration of the mechanical properties of composites, a combination of the mechanical properties of the constituent materials (fibre and matrix) [5] (Illustration: canva, 2024, https://www.canva.com).

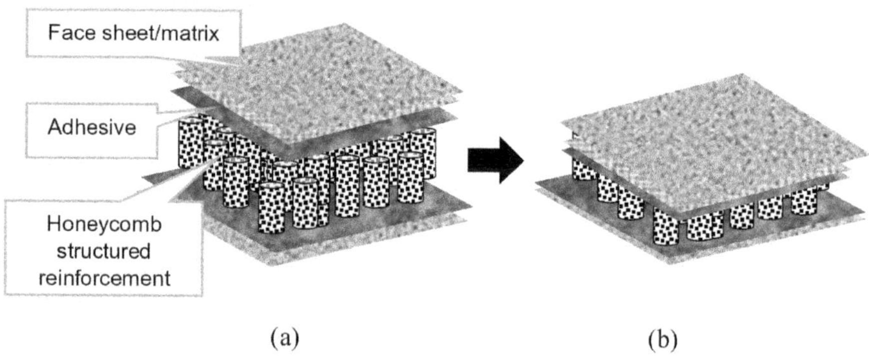

FIGURE 3.2
Typical composite elements with honeycomb structured reinforcement (a) and a fabricated sandwich panel (b) [6].

durability. This structural arrangement consists of hexagonal cells that create a strong and efficient load-bearing framework.

3.3 Matrix

The matrix in composite materials plays a crucial role in distributing loads, protecting the reinforcing fibres or particles, transferring loads, enhancing impact resistance, and maintaining dimensional stability. Its overall function is to provide strong and stable structural support for the composite, ensuring the desired mechanical, thermal, and environmental performance. Materials

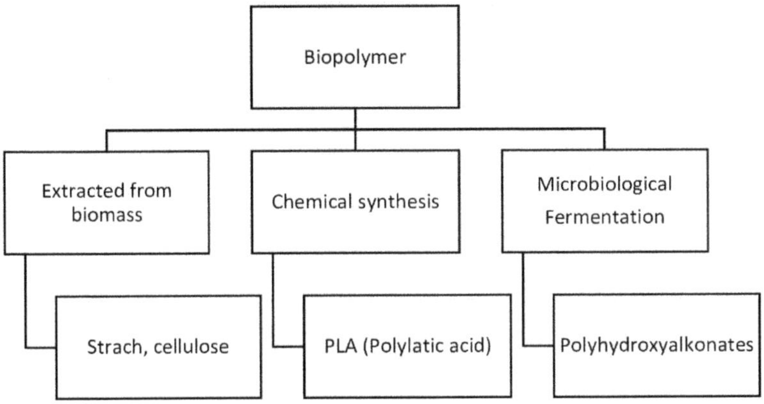

FIGURE 3.3
Biopolimer matrices classification [7].

that can be used as matrices include polymers such as epoxy, polyester, and polypropylene; metals such as aluminium, copper, and steel; and ceramics such as oxides and carbides. Matrices can also be made by natural materials such as the following.

3.3.1 *Biopolymer Matrix*

Biopolymer matrices can be classified based on their origin and composition. They are often categorized as either natural or synthetic, with each category having unique properties and applications (see **Figure 3.3**). Natural biopolymer matrices include materials derived from renewable resources, such as starch, cellulose, or chitosan, while synthetic biopolymer matrices are created through chemical processes and modifications.

3.3.2 *Biomass Matrix*

Biopolymers can be produced from biomass and organic materials. Biomass consists of materials such as wood, logging remnants, crops, and agricultural byproducts, and has been widely used in the generation of energy, fuels, and chemicals, serving as a promising renewable resource. Various parts of plants can serve as biomass, including their leaves, stems, roots, seeds, and even the whole plant. An important component of biomass plants is the cell wall, which includes main elements such as cellulose, hemicellulose, and lignin. Cellulose is a consistent feature in all lignocellulosic biomass, while the ratios and chemical compositions of lignin and hemicellulose can vary across different types of biomasses. **Figure 3.4** provides a simplified diagram illustrating the structure of the plant cell wall (cellulose, hemicellulose, and lignin). The arrangement and structure of these components within the cell wall, emphasize their roles in providing structural support, mechanical

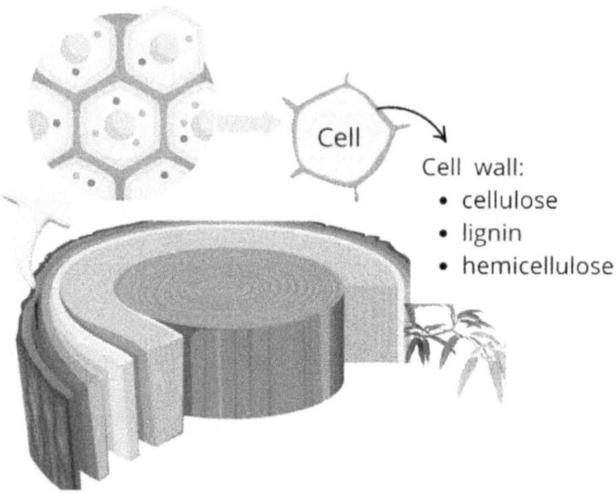

FIGURE 3.4
The schematic representation of the cell wall with the location of the main components in biomass [8] (Illustration: canva, 2024, https://www.canva.com).

strength, and functionality. The roles and functions of cellulose and lignin as materials for composite matrices are described below.

3.4 Cellulose-Based Materials

Cellulose is the predominant chemical constituent in lignocellulosic biomass, constituting roughly 50% by weight [8]. Cellulose is derived from a natural polysaccharide that is a major component of the cell walls of plants. Cellulose can also be extracted from certain bacteria species, various algae, and specific animals. Cellulose is one of the most abundant organic compounds on earth and is known for its strength, flexibility, and biodegradability. Cellulose-based materials can be produced from various sources, including wood, cotton, hemp, and other plant-based fibres.

Cellulose, a high-molecular-weight homobiopolymer, features a linear chain comprising numerous repeating D-glucose monomers linked by *β-1,4 glycosidic* bonds [9]. Within the cellulose chain, the hydroxyl groups establish connections through van der Waals forces and hydrogen bonds, creating both intramolecular and intermolecular bonds. These bonds facilitate the parallel assembly of elementary fibres, typically measuring 3–5 nm in width. These elementary fibres further combine to form microfibrils, characterized by widths of approximately 5–20 nm and lengths spanning a few micrometres [10]. Cellulose possesses diverse physical and chemical properties, making it a suitable matrix for creating a wide range of composites.

3.4.1 Classification of Cellulose

3.4.1.1 Cellulose Nanocrystals

Cellulose nanocrystals are tiny crystalline particles shaped like cylindrical rods, with diameters below 10 nm and lengths ranging from 100 nm to 500 nm. They are commonly produced by treating cellulose fibres with either acid hydrolysis or sulphuric acid. Sulphuric acid treatment results in nanocrystals with negatively charged sulphate groups on their surfaces, enhancing stability but reducing thermal stability and modifying reactivity. The properties of cellulose produced are influenced by acid concentration, hydrolysis temperature, and time [12]. Another approach involves enzymatic hydrolysis. This method is used to isolate cellulose nanocrystals, resulting in higher aspect ratios and greater thermal stability compared to those produced through acid hydrolysis. However, the nanocrystals generated using this method may exhibit lower colloidal stability due to their higher agglomeration tendency [13].

3.4.1.2 Cellulose Nanofibrils

Cellulose nanofibrils, characterized by their intricate network of nanoscale fibres with diameters ranging from 20 nm to 100 nm and lengths spanning several micrometres, have attracted considerable attention across various fields of research. These nanofibrils exhibit a combination of amorphous and crystalline domains, albeit with lower crystallinity compared to cellulose nanocrystals [14]. Employing diverse methods such as chemical, mechanical, and enzymatic treatments, either individually or in combination, facilitates the extraction of cellulose nanofibrils from lignocellulosic biomass. The commonly utilized TEMPO-mediated oxidation method introduces carboxyl groups at specific sites on the glucose units, resulting in nanofibrils of approximately 800 nm in length [15]. Alternatively, enzymatic pretreatment of wood pulp serves as a greener and cost-effective method for obtaining cellulose nanofibrils. Furthermore, ultrafine grinding represents another prevalent technique involving mechanical forces for the breakdown of cellulose into nano-sized fibrils.

3.4.1.3 Hairy Cellulose Nanocrystals

Hairy cellulose nanocrystals, a novel category of nanocellulose, feature a crystalline structure with amorphous cellulose chain protrusions at both ends. Their production involves subjecting cellulose fibrils to periodate oxidation to yield dialdehyde-modified cellulose, followed by heating and/or chemical treatment (chlorite oxidation or Schiff base reaction) to solubilize and selectively cleave the amorphous regions. This process can yield electrically neutral, negatively, or positively charged hairy cellulose nanocrystals [16].

3.5 Lignin-Based Materials

Lignin-based materials are derived from lignin, a complex organic polymer that is a key component of the cell walls of plants. Lignin is primarily found in the structural tissues of vascular plants and is responsible for providing rigidity and support [17]. Lignin comprises an amorphous macromolecule containing an aromatic structure made up of recurring phenylpropane units. it is the second most significant element of wood cell walls (20–30%). Manufacturing composites using lignin has been achieved through the incorporation of polymeric matrices such as *polypropylene, epoxy, polyvinyl alcohol, polylactic acid,* starch, wood fibre, natural rubber, and chitosan. Applications for these lignin-based composites have been observed in various fields, including packaging, biomedical materials, automotive components, advanced biocomposites, and flame retardant products [18]. Lignin can be utilized as a filler, stabilizer, compatibilizer, and reinforcement in composites due to its distinctive chemical structure. Lignin can be extracted using various methods, including the sulphur-free process: organosolv, soda; sulphur-bearing process: kraft, sulphite process, and the biorefinery process. These methods involve the separation of lignin from the other components of biomass, allowing for its isolation and use in downstream applications.

Lignin finds applications in various composite materials, including composite lignin-epoxy resin which offers improved mechanical properties and durability. Composite lignin polylactic acid presents an environmentally friendly option for packaging and biomedical purposes. Lignin-cellulose-based composites and lignin-polyvinyl alcohol composites provide excellent film-forming capabilities. The development of lignin-starch composites is particularly valuable in the effort to reduce plastic waste. Furthermore, lignin-chitosan composites exhibit unique properties, such as antimicrobial and wound-healing capabilities. Lastly, lignin-based rubber composites contribute to the enhancement of rubber materials, improving their mechanical strength and thermal stability [19].

3.6 Polylactic Acid (PLA)

PLA is a biodegradable and bioactive thermoplastic derived from renewable resources, such as corn starch, tapioca roots, or sugarcane. It is considered a

TABLE 3.2

Physical Properties of PLA and PLLA [21]

Bio polymers	Density (g/cm³)	Tensile Strength (MPa)	Youngs Modulus (GPa)	Glass Transition Temperature (°C)	Melting Temperature (°C)
PLA	1.21–1.25	21–60	0.35–3.5	45–60	150–162
PLLA	1.24–1.30	15.5–150	2.7–4.14	55–65	170–200

PLA: *polylactic acid*
PLLA: *poly-l-lactic acid*

sustainable and environmentally friendly alternative to conventional petroleum-based plastics. PLA is commonly mixed with various other materials to achieve the desired properties [20]. It is often mixed with carbon fibres and conductive carbon. In the form of PLA filaments, it can also be made into coloured fibres. The mechanical properties of the PLA matrix vary depending on factors such as temperature, processing techniques, and the specific composition of the material. **Table 3.2** displays the mechanical properties of PLA such as strength, flexibility, density, melting temperature, and glass transition temperature.

3.6.1 Natural Rubber

Natural rubber, also known as latex, is a type of elastomer derived from the milky sap of various plants, with *Hevea brasiliensis* being the primary source. It is composed primarily of polymers of isoprene, a volatile liquid monomer, which can be coagulated and solidified into a resilient and elastic material through various processes.

Table 3.3 shows the inherent physical properties of natural rubber. Those inherent properties of natural rubber limit its applications, which need the addition of compounds to meet various requirements. This involves incorporating other materials that enhance its properties, modifying its characteristics, reducing costs, and extending its lifespan. Among these additives are fillers, *zinc oxide,* processing oil, *tetramethyl thiuram disulphide* (TMTD),

TABLE 3.3

Inherent Physical Properties of Natural Rubber [22]

Parameter Value	Value
Density at 20°C	0.906–0.916 g/cm³
Specific heat	1.905 $KJg^{-1}K^{-1}$
Refractive index	1.5191
Heat of combustion	45.2KJ/Kg

mercaptobenzothiazole sulphanamide (MBTS), *trimethyl quinoline* (TMQ), *stearic acid, carbon black* (N330 HAF), and *sulphur* [22].

3.7 Natural Fibre-reinforced PMCs

The advancement of polymer composites (PMC) based on natural fibres has initiated to solve environmental problems. The robust mechanical properties and ability to resist hydrolysis of natural fibres can be attributed to the presence of cellulose and hemicellulose within their composition. The presence of organic or inorganic substances can also be beneficial for the natural fibre's odour, colour, breakdown resistance, and properties. Typical mechanical properties of commonly used natural fibres are listed in **Table 3.4** [23].

Natural fibres are commonly sourced from minerals or animals. However, plant-based fibres are preferred for use in the production of composite materials. Unlike fibres obtained from animals like natural silk, plant-based fibres utilize sustainable additives from production to application, rendering them more cost-effective than animal-based fibres. Animal- or mineral-based fibres still necessitate the addition of synthetic materials during processing. Some common natural fibres include: *cotton, jute, hemp, flax, sisal, coir, wool, silk, ramie,* and *bamboo.* These natural fibres are widely utilized in various industries for

TABLE 3.4

Typical Mechanical Properties of Natural Fibres Reinforced Polymer Composites [23]

Composites		Mechanical Properties		
Polymer Matrix	Natural Fibre	Tensile Strength (MPa)	Tensile Modulus (Gpa)	Flexural Strength
JC-02A epoxy resin	Raw jute	79		
Polylactic acid	Raw jute	11.9	0.2	
L12 epoxy resin	Gsm 210 jute	33		46
Green epoxy resin	Bast flax	25.87	0.494	
LY556 epoxy resin	Bast flax	45		112.5
1010 polyamide	Hemp	58	2.9	
DER331 epoxy resin	Temafa kenaf	159.98	8.20	
INF114 epoxy resin	Kenaf	76.67	2.31	61.24
Bisphenol-A epoxy resin	Curaua	134.67	3.08	
LY556 epoxy resin	Bamboo	40.28		
Bisphenol-A epoxy resin	Bamboo	111.54	3.9	45.23

their versatility, sustainability, and biodegradability. They find applications in textiles, construction, and composite materials, among other areas.

These composites offer various advantages, including lower specific weight, high specific strength, and stiffness. Additionally, they are abundantly available, biodegradable, and exhibit low abrasion, reducing tool wear during production. However, natural fibre composites also have their limitations, such as relatively low mechanical properties, moisture sensitivity, and poor fire resistance. Overcoming these challenges is crucial for their widespread application in various industries. **Figure 3.5** presents sources of natural fibre which can be used for composite reinforcement [24].

Several natural fibres, such as coconut, coir, cotton, flax, hemp, and jute, demonstrate a wide range of physical and mechanical properties, making them valuable candidates for various applications. **Table 3.5** provides a comprehensive overview of these properties, highlighting their potential uses in industries such as textiles, construction, and manufacturing.

3.7.1 Lignin as Filler in Wood-Based Composite

Lignin plays a crucial role as a substitute for synthetic materials in composite adhesives. It possesses a complex structure, low reactivity, low solubility, high polydispersity, and strong hydrophobicity. Because of its limited chemical reactivity, lignin necessitates high catalyst concentrations and longer processing times [27]. Modification is essential to enhance lignin properties, achieved through the incorporation of additional substances like starch,

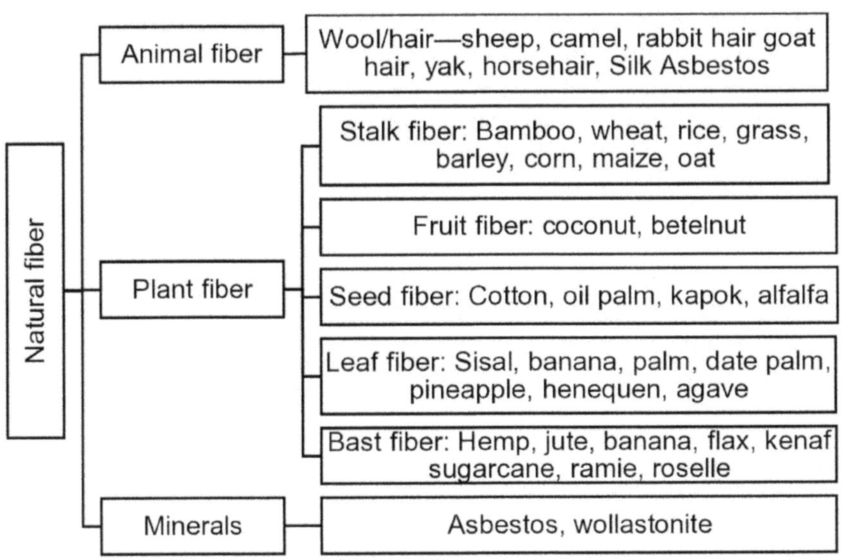

FIGURE 3.5
Classification of natural fibres adapted from [7, 25]

TABLE 3.5
Physical and Mechanical Properties of Natural Fibres [26]

Fibre	Density (g/cm³)	Tensile Strength (MPa)	Young Modulus (GPa)	Elongation (%)
Coconut	1.15	131–175	4–6	15
Coir	1.2	175–220	4–6	15–30
Cotton	1.5–1.6	287–597	5.5–12.6	3–10
Flax	1.4–1.5	345–1500	27.6–80	1.2–3.2
Hemp	1.48	550–900	70	1.6
Jute	1.3–1.46	393–800	10–30	1.5–1.8
Pineapple leaf	1.07–1.50	413–1627	34.5–82.5	
Ramie	1.5	220–938	44–128	2.0–3.8
Sisal	1.33–1.5	400–700	9–38	2–14
Abaca	1.5	410–810	41	1.6
Chicken feathers	0.89	100–200	3–10	–

tannin, polyethyleneimine, or glyoxal [28]. Various types of lignin, including KL (kraft lignin), OL (Organosolv lignin), SL (soda lignin), or hydrolysis lignin [29], have found applications in plywood adhesives, oriented strand board (OSB), fibreboard, particleboard, and structural material applications.

One of the lignin additive materials is ammonium LS. It has a high density and can be used for fibreboard by mixing it with UF resin in the amount of 4–8% of the wood weight. The use of 6% ammonium LS can reduce the mechanical properties of wood fibres due to the pressure factor and increase moisture content. Magnesium lignosulphonate is commonly used as a binder for fibreboards and certain structural materials with a 15% adhesive content by weight [30]. Lignin can also be utilized as a substitute material for phenol at levels below 50% [31]. Additional chemicals such as formaldehyde and phenol are employed in adhesives derived from lignin. Enzymatically hydrolyzed lignin was combined with NR latex to produce a bio-adhesive using a blending method with a filler as the base component, suitable for medium-density fibreboard adhesion. The inclusion of 5 and 10 g of lignin into 10 g of natural rubber latex resulted in the most elevated values for the modulus of elasticity and modulus of rupture for medium-density fibreboard, as reported in reference [32].

3.8 Plant Fibres/Fillers

3.8.1 Rice Husk Ash (RHA)

Rice husk ash (RHA) is a commonly available waste product produced from the burning of rice husks. During the burning process, the majority of the

TABLE 3.6

Mechanical Properties of Composites with RHA as a Fibre Filler [33]

Samples	Flexure Strain (%)	Flexure Stress at Tensile (MPa)	Young's Modulus (GPa)	Flexure Displacement (mm)	Time at Tensile
5% Fine	1.52	60.144	4.334	1.67	45.87
10% Fine	2.674	79.906	4.02	2.936	91.31
15% Fine	1.67	20.17	1.626	1.832	64.75
20% Fine	1.196	61.198	5.876	1.312	36.86

evaporable components of rice husk are progressively destroyed, leaving just the silicates as the principal leftovers. The main composition of rice hush consists of silicates and a significant amount of organic carbon. Its properties are influenced by the combustion process, including variables such as temperature and duration. For every 100 kg of husks burnt in a boiler, 25 kg of RHA is produced. Through controlled combustion of rice husks over an extended period, the volatile organic matter, largely comprising cellulose and lignin, is removed, leaving ash predominantly composed of amorphous silica with a microporous cellular structure [33]. RHA exhibits a high specific surface area, ranging from 20 to 270 m^2/g, making it a potential material for various applications. Rice husk ash has the potential to act as a filler, enhancing the properties of composites and making it a viable option for eco-friendly materials [34]. The effects of varying concentrations of rice husk on mechanical properties are presented in the **Table 3.6.**

3.8.2 Pineapple Leaf Fibre (PALF)

The pineapple leaf fibre (PALF) is a biofibre that demonstrates biodegradable and environmentally friendly properties, along with good tensile strength. The mechanical properties of PALF are presented in **Table 3.7**. The study utilized natural PALF with weight ratios of 0%, 5%, 10%, and 15%, revealing that the tensile strength (TS) and Young's modulus reached the highest levels for untreated 10% PALF, while the impact and flexural properties exhibited a decrease with the increasing fibre content. The addition of MAPE (*maleic anhydride polyethylene*) resulted in lower tensile properties compared

TABLE 3.7

Physical and Mechanical Strength of PALF [37]

Density (g/cm³)	Tensile Strength (MPa)	Young's Modulus (GPa)	Specific Modulus (GPa/g/cm³)	Elongation at Break (%)	Dia. (μm)	PALF Fibre/ Reference
1.52	413–1627	34.5–82.51	22.7–54.3	1.6–3	20–80	[38]

to the composite without MAPE treatment. However, the presence of MAPE enhanced the composite's flexural and impact properties[35].

Moreover, the strength of the composite is influenced by the modification of the fibre surface through various alkaline treatments. Increasing the alkaline concentration improved the tensile strength of the PALF composite and its tensile-to-break ratio (TBR) by up to 20%, due to enhanced surface roughness and increased reaction exposure. However, the highest alkaline concentration led to composite weakening due to fibre damage. A 5% alkaline solution is deemed the most effective for NaOH treatment. The highest flexural strength, recorded at 5% alkaline treatment, was 6.37 MPa, marking an 11.1% increase from the 3% concentration, which yielded 5.73 MPa [36].

3.8.3 Jutes

Jute is a natural fibre obtained from the stem of the jute plant commonly known as lignocellulosic fibres [39]. They are commonly used in the production of textiles, particularly in the manufacturing of burlap, rope, twine, and sacks. Ahmed and colleagues [40] conducted a series of tests on the effects of glass fibre when added to jute fibre. The research indicated that the tensile strength, interlaminar shear, and flexural strength properties showed a significant improvement with the addition of glass fibre. The use of 16.5% by weight of glass fibre increased the shear, tensile, and flexural strength properties by 17.6%, 37%, and 31.23%, respectively. To enhance the mechanical properties, surface treatments of jute-based composites are employed.

The effects of surface treatments on enhancing the mechanical properties of jute fibres have been extensively studied by researchers [41–43]. NaOH treatment is capable of altering the physical, chemical, and physicochemical surfaces of the composites. Jute/polylactide composites comprising 50% jute treated with a 5% aqueous NaOH solution exhibited increased roughness and improved fibre/matrix adhesion [42]. *Jute/polypropylene* composites with 25% jute treated with urea demonstrated a substantial increase in tensile strength [43]. Jute/epoxy manufactured using the hand lay-up technique and treated with 20% and 7% NaOH also showed enhanced tensile strength [44].

3.8.4 Kenaf Fibres

Kenaf fibre (KF) is a natural bast fibre derived from the kenaf plant (*Hibiscus cannabinus*), a member of the *Malvaceae* family [45]. Although native to Africa, it is now cultivated worldwide for its versatile and sustainable fibre. Kenaf fibre finds common use in the production of textiles, ropes, cordage, and other materials requiring strong and durable fibres. The kenaf fibres are obtained from the stalk of the plant, comprising two primary components. Approximately 35% of the plant's total weight is comprised of the outer part, which contains long fibres known as bast fibre. The inner part, known as the core, is a wooden component containing short fibres [46]. Kenaf fibres can be enhanced in their mechanical strength through various treatments.

The improvement occurs when impurities and some water-absorbing chemical components such as lignin, hemicellulose, wax, and oils from the outer surface of the fibres are removed. Through surface roughness treatment, this can increase the bonding with the matrix.

Several studies have been conducted on the use of kenaf as a composite material, with fibre volume fractions ranging from 0.25 to 2 vol.% [47]. On average, researchers use kenaf mixtures ranging from 0.1 to 2.5% of vol., although few have investigated the impact of KF content above 2.5% of vol. The length of kenaf fibres used in these studies ranged from 10 mm to 80 mm. Abirami, Guo et al., and Beddu [48–50] have conducted experiments utilizing kenaf fibre as reinforcement, with percentages ranging from 0.2% to 1%, and all of the results confirmed that kenaf fibre enhances the tensile strength of the composite. The use of an alkaline solution has been found to improve the compressive and flexural strengths. Baarimah [51] observed this by using three different concentrations (1%, 3%, and 6%) of the compositions and two distinct volume fractions of KF (1% and 2%). The composites containing 1% KF treated with a 6% concentration of NaOH solution showed the highest compressive and flexural strengths.

References

1. Visakh, P. M., Bayraktar, O., & Menon, G. (2019). *Bio monomers for green polymeric composite materials*. Willey.
2. Chen, J., Nie, X., Liu, Z., Mi, Z., & Zhou, Y. (2015). Synthesis and application of polyepoxide cardanol glycidyl ether as biobased polyepoxide reactive diluent for epoxy resin. *ACS Sustainable Chemistry & Engineering, 3*(6), 1164–1171.
3. Kumar, K., & Sekaran, A. (2014). Some natural fibers used in polymer composites and their extraction processes: A review. *Journal of Reinforced Plastics and Composites, 33*(20), 1879–1892.
4. Egute, N. S., Forster, P. L., Parra, D. F., Fermino, D. M., Santana, S., & Lugao, A. B. (2009). *Mechanical and thermal properties of polypropylene composites with curaua fibre irradiated with gamma radiation*. Proceedings of the International Nuclear Atlantic Conference.
5. Callister, W. D., Jr. (2018). *Materials science and engineering: An introduction* (10th ed.). Wiley. https://www.canva.com
6. Ashby, M. (2010). *Materials selection in mechanical design*. Butterworth-Heinemann.
7. Shekar, H. S., & Ramachandra, M. (2018). Green composites: A review. *Materials Today: Proceedings, 5*, 2518–2526.
8. Shen, D., Xiao, R., Gu, S., & Zhang, H. (2013). *The overview of thermal decomposition of cellulose in lignocellulosic biomass*. InTech. doi:10.5772/51883. https://www.canva.com
9. Sampath, U., Ching, Y. C., Chuah, C. H., Sabariah, J. J., & Lin, P. C. (2016). Fabrication of porous materials from natural/synthetic biopolymers and their composites. *Materials, 9*, 991.

10. Ha, M. A., Apperley, D. C., Evans, B. W., Huxham, I. M., Jardine, W. G., Vietor, R. J., Reis, D., Vian, B., & Jarvis, M. C. (1998). Fine structure in cellulose microfibrils: NMR evidence from onion and quince. *Plant Journal, 16*, 183–190.
11. Teng, C. P., Tan, M. Y., Toh, J. P. W., Lim, Q. F., Wang, X., Ponsford, D., Lin, E. M. J., Thitsartarn, W., & Tee, S. Y. (2023). Advances in cellulose-based composites for energy applications. *Materials, 16*(10), 3856. https://doi.org/10.3390/ma16103856
12. Chen, L., Wang, Q., Hirth, K., Baez, C., Agarwal, U. P., & Zhu, J. Y. (2015). Tailoring the yield and characteristics of wood cellulose nanocrystals (CNC) using concentrated acid hydrolysis. *Cellulose, 22*, 1753–1762.
13. Sacui, I. A., Nieuwendaal, R. C., Burnett, D. J., Stranick, S. J., Jorfi, M., Weder, C., Foster, E. J., Olsson, R. T., & Gilman, J. W. (2014). Comparison of the properties of cellulose nanocrystals and cellulose nanofibrils isolated from bacteria, tunicate, and wood processed using acid, enzymatic, mechanical, and oxidative methods. *ACS Applied Materials & Interfaces, 6*, 6127–6138.
14. Trache, D., Tarchoun, A. F., Derradji, M., Hamidon, T. S., Masruchin, N., Brosse, N., & Hussin, M. H. (2020). Nanocellulose: From fundamentals to advanced applications. *Frontiers in Chemistry, 8*, 392.
15. Yang, X., Reid, M. S., Olsen, P., & Berglund, L. A. (2020). Eco-friendly cellulose nanofibrils designed by nature: Effects from preserving native state. *ACS Nano, 14*, 724–735.
16. Muthami, J., Wamea, P., Pitcher, M., Sakib, M. N., Liu, Z., Arora, S., Kennedy, D., Chang, Y.-J., & Sheikhi, A. (2021). Hairy cellulose nanocrystals: From synthesis to advanced applications in the water–energy–health–food nexus. In cellulose nanoparticles: Volume 2: Synthesis and manufacturing. *The Royal Society of Chemistry*, 2, 1–37.
17. Balk, M., Sofia, P., Neffe, A. T., & Tirelli, N. (2023). Lignin, the lignification process, and advanced, lignin-based materials. *International Journal of Molecular Sciences, 24*(14), 11668. https://doi.org/10.3390/ijms241411668
18. Ridho, M. R., Agustiany, E. A., Rahmi, M., et al. (2022). Lignin as green filler in polymer composites: Development methods, characteristics, and potential applications. *Advances in Materials Science and Engineering, 2022*, Article ID 1363481, 33 pages. https://doi.org/10.1155/2022/1363481
19. Ma, C., Kim, T.-H., Liu, K., Ma, M.-G., Choi, S.-E., & Si, C. (2021). Multifunctional lignin-based composite materials for emerging applications. *Frontiers in Bioengineering and Biotechnology, 9, 708976*. https://doi.org/10.3389/fbioe.2021.708976
20. Sanivada, U. K., Mármol, G., Brito, F. P., & Fangueiro, R. (2020). PLA composites reinforced with flax and jute fibers—a review of recent trends, processing parameters and mechanical properties. *Polymers, 12*(10), 2373. https://doi.org/10.3390/polym12102373
21. Van De Velde, K., & Kiekens, P. (2001). Biopolymers: Overview of several properties and consequences on their applications. *Polymer Testing, 99*, 483.
22. Aguele, F. O., & Madufor, C. I. (2012). Effects of carbonised coir on physical properties of natural rubber composites. *American Journal of Polymer Science*, 2(3), 28–34. https://doi.org/10.5923/j.ajps.20120203.02

23. Sharma, H., Kumar, A., Rana, S., Sahoo, N. G., Jamil, M., Kumar, R., Sharma, S., Li, C., Kumar, A., Eldin, S. M., & Abbas, M. (2023). Critical review on advancements on the fiber-reinforced composites: Role of fiber/matrix modification on the performance of the fibrous composites. *Journal of Materials Research and Technology, 26,* 2975–3002.
24. Jawaid, M., & Khalil, H. A. (2011). Cellulosic/synthetic fiber reinforced polymer hybrid composites: A review. *Carbohydrate Polymers, 86,* 1–18.
25. Sanivada, U. K., Mármol, G., Brito, F. P., & Fangueiro, R. (2020). PLA composites reinforced with flax and jute fibers—a review of recent trends, processing parameters and mechanical properties. *Polymers, 12*(10), 2373.
26. Siakeng, R., Jawaid, M., Ariffin, H., Sapuan, S. M., Asim, M., & Saba, N. (2018). Natural fiber reinforced polylactic acid composites: A review. *Polymer Composites, 40,* 446–463.
27. Vishtal, A., & Kraslawski, A. (2011). Challenges in industrial applications of technical lignins. *Bio, 6*(3), 3547–3568.
28. Petar, S., Viktor, S., & Nikolay, N. (2020). Sustainable bio-based adhesives for eco-friendly wood composites: A review. *Wood Research, 65*(1), 51–62.
29. Antov, P., Savov, V., Trichkov, N., et al. (2021). Properties of high-density fiberboard bonded with urea–formaldehyde resin and ammonium lignosulfonate as a bio-based additive. *Polymers, 13*(16), 2775.
30. Antov, P., Jivkov, V., Savov, V., Simeonova, R., & Yavorov, N. (2020). Structural application of eco-friendly composites from recycled wood fibres bonded with magnesium lignosulfonate. *Applied Sciences, 10*(21), 7526.
31. Danielson & Simonson, R. (1998). Kraft lignin in phenol formaldehyde resin. Part 1. Partial replacement of phenol by kraft lignin in phenol formaldehyde adhesives for plywood. *Journal of Adhesion Science and Technology, 12*(9), 923–939.
32. Thuraisingam, J., Mishra, P., Gupta, A., Soubam, T., & Piah, B. M. (2019). Novel natural rubber latex/lignin-based bio-adhesive: Synthesis and its application on medium density fiber-board. *Iranian Polymer Journal* (English Edition), *28*(4), 283–290.
33. Shehab, Rice hush ash for composites, Final Year Project 2021. International Univsersity.
34. Agrela, F., Cabrera, M., Morales, M., Zamorano, M., & Alshaaer, M., 2021. Biomass fly ash and biomass bottom ash. In *New trends in eco-efficient and recycled concrete* (pp. 23–58). Woodhead Publishing. https://doi.org/10.1016/B978-0-08-102480-5.00002-6
35. Siregar, J. P., Jaafar, J., Cionita, T., et al. (2019). The effect of maleic anhydride polyethylene on mechanical properties of pineapple leaf fibre reinforced polylactic acid composites. *International Journal of Precision Engineering and Manufacturing-Green Technology, 6,* 101–112.
36. Mathivanan, D. B., Siregar, J. P., Mat Rejab, M. R., Bachtiar, D., Asmara, Y. P., & Cionita, T. (2017). The mechanical properties of alkaline treated pineapple leaf fibre to reinforce tapioca based bioplastic resin composite. *Materials Science Forum, 882,* 66–70.
37. Asim, M., Abdan, K., Jawaid, M., Nasir, M., Dashtizadeh, Z., Ishak, M. R., & Hoque, M. E. (2015). A review on pineapple leaves fibre and its composites. *International Journal of Polymer Science, 2015,* Article ID 950567, 16 pages.
38. Fakirov, D. B. S. (2007). *Wood fibre thermoplastic composite: Processing properties and future developments.* Hanser.

39. Chaudhary, V., Bajpai, P. K., & Maheshwari, S. (2018). Studies on mechanical and morphological characterization of developed Jute/Hemp/Flax reinforced hybrid composites for structural applications. *Journal of Natural Fibers, 15,* 80–97.
40. Ahmed, K. S., Vijayarangan, S., & Rajput, C. (2006). Mechanical behavior of isothalic polyester-based untreated woven jute and glass fabric hybrid composites. *Journal of Reinforced Plastics and Composites, 25,* 1549–1569.
41. Ashraf, M. A., Zwawi, M., Mehran, M. T., Kanthasamy, R., & Bahadar, A. (2019). Jute based bio and hybrid composites and their applications. *Fibers, 7*(9), 77. https://doi.org/10.3390/fib7090077
42. Goriparthi, B. K., Suman, K., & Rao, N. M. (2012). Effect of fiber surface treatments on mechanical and abrasive wear performance of polylactide/jute composites. *Composites Part A: Applied Science and Manufacturing, 43,* 1800–1808.
43. Rahman, R., Hasan, M., Huque, M. M., & Islam, M. N. (2010). Physico-mechanical properties of jute fiber reinforced polypropylene composites. *Journal of Reinforced Plastics and Composites, 29,* 445–455.
44. Boopalan, M., Umapathy, M. J., & Jenyfer, P. (2012). A comparative study on the mechanical properties of jute and sisal fiber reinforced polymer composites. *Silicon, 4,* 145–149.
45. Abbas, A.-G. N., Aziz, F. N. A. A., Abdan, K., & Mohd Nasir, N. A. (2022). Kenaf fibre reinforced cementitious composites. *Fibers, 10*(1), 3. https://doi.org/10.3390/fib10010003
46. Juliana, A. H., Paridah, M. T., Rahim, S., Azowa, I. N., & Anwar, U. M. K. (2012). Properties of particleboard made from kenaf (Hibiscus cannabinus L.) as function of particle geometry. *Journal of Materials, 34,* 406–411.
47. Lande, I., & Terje, R. (2020). The influence of steel fibres on compressive and tensile strength of ultra high performance concrete: A review. *Construction and Building Materials, 256,* 119459.
48. Guo, A., Sun, Z., & Satyavolu, J. (2021). Experimental and finite element analysis on flexural behavior of mortar beams with chemically modified kenaf fibers. *Construction and Building Materials, 292,* 123449.
49. Beddu, S., Basri, A., Muda, Z. C., Farahlina, F., Mohamad, D., Itam, Z., Kamal, N. L. M., & Sabariah, T. (2021). Comparison of thermomechanical properties of cement mortar with kenaf and polypropylene fibers. *IOP Conference Series: Materials Science and Engineering, 1144,* 012036.
50. Abirami, R., Sangeetha, S. P., Nadeemmishab, K., Vaseem, P. Y., & Sad, K. S. (2020). Experimental behaviour of sisal and kenaf fibre reinforced concrete. *AIP Conference Proceedings, 2271,* 030023.
51. Baarimah, A. O., Mohsin, S. M. S., Alaloul, W. S., & Ba-Naimoon, M. S. (2021). Effect of sodium hydroxide on mechanical characteristics of kenaf fibers reinforced concrete. *Journal of Physics: Conference Series,* 1962, 012013.

4

Green Concrete for Sustainable Infrastructure Development

4.1 Introduction

The rapid development of infrastructure and the expansion of new areas are driving an increasing demand for cement. The fast-growing industrialization of cement production poses a significant threat to the environment, leading to decreased air quality, heightened water waste, and soil degradation. This rise in cement production contributes substantially to CO_2 emissions from concrete, which range 5–8% CO_2 emission [1]. It is also recorded that the production of Portland clinker emits approximately 1 kg of CO_2 into the atmosphere for every 1 kg of clinker produced [2]. Furthermore, this industrial growth contributes to the depletion of natural resources, the deterioration of environmental quality, and the challenges associated with managing increasing amounts of waste. Considering those issues, green concrete marks a recent breakthrough in the construction industry, emphasizing the use of sustainable materials and methods. This innovative approach integrates various recycled materials, including fly ash, silica fume, slag (FA), palm oil fuel ash (POFA), kaolin, metakaolin, dolomite, and carbon nanotube to achieve superior properties [3–5]. These materials not only reduce environmental impacts but also promote the efficient management of industrial by-products. This innovation seeks to create concrete that is not only robust and long-lasting but also eco-friendly. Through the adoption of sustainable technologies and alternative materials, green concrete has become increasingly popular, offering a more environmentally friendly solution for modern, sustainable infrastructure. The global green cement market is projected to grow at a compound annual growth rate (CAGR) of 14.1% from 2017 to 2023 [4]. The potential of using alternative materials has been found to reduce carbon emissions from concrete by up to 33% [6].

Green concrete minimizes the use of cement by incorporating biodegradable materials or waste-based additives. It involves partially or entirely replacing ordinary Portland cement (OPC) with alternative materials and using waste and recycled materials as aggregates to improve various concrete

DOI: 10.1201/9781003617594-4

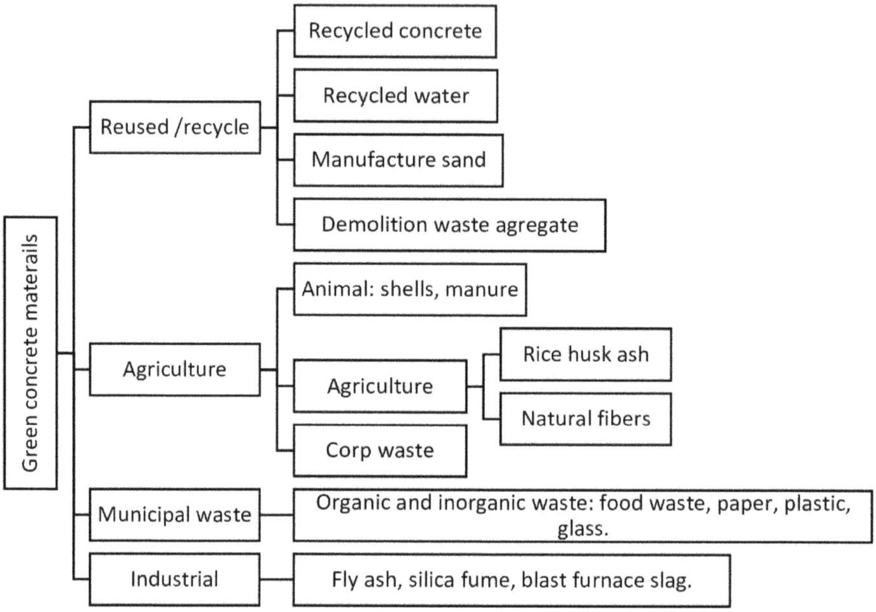

FIGURE 4.1
Materials used for green concrete [4, 5].

properties such as workability, tensile strength, compression, and durability [7]. Recycled waste materials suitable for concrete production are typically categorized into three groups: industrial, agricultural, and municipal waste, as presented in **Figure 4.1**.

Prior to their incorporation into concrete, these waste materials undergo physical, chemical, or physicochemical processing. This process aims to break down the materials into smaller particles, thereby increasing their surface area and reactivity. To further boost reactivity, chemical substances like sodium sulphate anhydrite, sodium silicate, acid, or calcium formate are employed, activating the pozzolanic properties of the cementitious materials. These activation methods are crucial for minimizing inconsistencies in their chemical characteristics. More details regarding materials commonly used in the production of green concrete include the following:

- *Fly ash*: a byproduct of coal-fired power plants, and is used as a partial replacement for Portland cement in concrete.
- *Silica fume*: a byproduct of the production of silicon and ferrosilicon alloys, and is used as a cementitious material in concrete.
- *Slag*: a byproduct of the metallurgical industry, and is used as a supplementary cementitious material in concrete.

- *Recycled aggregates*: aggregates sourced from recycled materials, such as crushed concrete or reclaimed asphalt pavement.
- *Ground granulated blast-furnace slag (GGBFS)*: a byproduct of the iron-making process which is used to replace cement in concrete production.
- *Geopolymers*: inorganic polymers made from industrial byproducts, often used as an alternative binder in concrete production.
- *Rice husk ash*: an agricultural waste product used as a supplementary cementitious material.
- *Nanomaterials*: nanoscale materials such as nanosilica and carbon nanotubes are used to improve the properties of concrete.
- *Recycled water*: treated wastewater or greywater used in the mixing and curing process of concrete, reducing water consumption.
- *Natural fibres*: organic materials such as jute, hemp, or kenaf are used as reinforcement in concrete, reducing the need for traditional steel reinforcement.

Figure 4.2 presents the arrangement of reinforced steel bars used in building construction. This arrangement demonstrates the placement and orientation of steel bars within concrete structures to enhance tensile strength and durability.

4.1.1 Fly Ash

Fly ash is a fine, powdery residue derived from the combustion of pulverized coal in coal-fired power plants. It consists primarily of spherical particles rich in *silicon dioxide* (SiO_2), *aluminium oxide* (Al_2O_3), *and iron oxide* (Fe_2O_3) [8].

FIGURE 4.2
Typical arrangement of reinforced steel bar used for construction building (Illustration: canva, 2024, https://www.canva.com).

The composition of fly ash varies depending on the type of coal burnt and combustion conditions. More than 80% of the coal ash produced is classified as fly ash (FA), while the remainder is bottom ash (BA) [9]. Fly ash poses serious environmental concerns due to its potential to leach heavy metals and other contaminants into soil and water if not managed properly. However, when utilized in concrete and other construction materials, it contributes to environmental sustainability by reducing the need for cement and enhancing structural durability. Currently, the management of fly ash involves either placement in landfill areas or storage in specific locations, requiring significant land use [10]. This process includes capturing it through water flows from boiler bottoms and collecting it via electrostatic or mechanical precipitation from exhaust gases.

Fly ash can be repurposed as a beneficial material, such as an additive in concrete to reduce cement content. It has been demonstrated that high fly ash content can be effectively used in concrete mixes for both normal and aggressive exposure conditions. Concrete with high fly ash content, known as green concrete, can be effectively used in both normal and aggressive exposure conditions. In the study [11], different mixtures were analyzed by using concrete mixture composition (cement: micro silica: fly ash: 57:5:38 with a water/binder ratio of 0.42) exposed to aggressive environments. The results indicated that the quality of green concrete achieved 40–60 MPa, with better workability and durability compared to conventional concrete. High volume fly ash (HVFA) concrete, which substitutes up to 50% of cement by mass with fly ash, has demonstrated significant improvements in mechanical properties and implementation principles.

Substituting a portion of cement with fly ash (FA) in concrete primarily serves as a filler and enhances pozzolanic activity. The incorporation of FA reduces the void ratio of Portland cement (PC) concrete by creating a secondary *C–S–H* gel, which improves cohesive activity and decreases voids and permeability. Research indicates that the void content and permeability in FA-enhanced concrete range from 13 to 35% and 0.03–1.3 cm/s, respectively [8]. For example, experiments replacing 10–20% of cement with Class F FA have shown a reduction of total voids by about 12–16% compared to the control mix [12]. *Muthaiyan and Thirumalai* (2017) [13] recorded FA permeability ranging from 1.19 cm/s to 0.641 cm/s when FA was added. The resulting low-porosity concrete offers numerous benefits, such as enhanced freeze-thaw resistance and impermeability, minimizing the risk of cracking and water infiltration. This is especially important for structures exposed to water, such as dams, bridges, and underwater foundations. Low-porosity concrete also lowers the risk of corrosion in reinforced concrete by limiting moisture and oxygen access to the steel reinforcement. Additionally, it results in a smoother surface finish, reducing the need for extra treatments, and lowers maintenance costs due to its enhanced durability. Moreover, it increases resistance to chemical attacks, enhancing durability and making it suitable for industrial environments [14].

Concrete with high-volume natural pozzolan (50% by mass of cementitious material), or high-volume natural pozzolan (HVNP), can achieve strengths of 14 MPa (at three days) and 38 MPa (at 28 days). This study utilizes natural pozzolans such as *low-calcium fly ash* and *high-lime furnace slag* [15]. Higher proportions of fly ash can be used as a substitute for cement, thereby reducing cement content. This concrete can serve as high-volume fly ash concrete in earthquake-resistant building construction [16]. In further applications, 100% fly ash concrete mixed with conventional aggregates and recycled pulverized glass aggregates has been extensively studied [17]. These mixtures exhibit workability (10 to 15 cm slump) and strength (at least 27.6 MPa at 28 days) comparable to standard concrete used in common construction practices. Both mixes have demonstrated strengths exceeding 55.2 MPa at 84 days, with a modulus of elasticity of 25.3 GPa, similar to conventional concrete. The concrete also performed well in terms of alkali silica reactivity. Several large-scale pilot projects over the past decade have applied 100% fly ash concrete in various structural applications (such as foundations, footings, floor slab, and beams for office buildings) and non-structural applications (including architectural panels), showing satisfactory results.

Jayanta 2016, [18] designed concrete mixes for M25, M35, and M50 grades using varying percentages of fly ash: 15%, 25%, 35%, 45%, 55%, and 65% by weight. The concrete cubes were cured for 7, 28, and 60 days to evaluate their compressive strength over time. Generally, as the proportion of fly ash increases in concrete mixes, experiments conducted over various durations indicate a decrease in compressive strength. This variability highlights the importance of carefully selecting and optimizing concrete mixes based on specific project requirements and performance expectations. It is recommended that the proportion of fly ash in the mix should not exceed 45% to maintain adequate strength properties. Substituting cement with fly ash (FA) at levels of 10–30% is optimal. Increasing the replacement level beyond this optimum range (30–50%) can adversely affect the hydration process, leading to slower strength gain and negatively impacting the overall strength of the concrete [19]. Liu's study [20] demonstrated similar findings when incorporating PC with 3%, 6%, 9%, and 12% FA, indicating that higher FA content led to decreased flexural and compressive strength during the early curing period (28 days). It was observed that with an increase in FA content up to 12%, compressive strength decreased by 34%. It may be related to the reduction in cement, which primarily results in a significant decrease in the cement paste used to coat the coarse aggregate and diminishes the bond between aggregates [19].

4.1.2 Waste Incineration Ash

Waste incineration ash is the residual material left after burning municipal solid waste (MSW) or other waste materials in specialized waste-to-energy plants. High temperatures in these plants are used to significantly reduce waste volume and generate energy [21]. This ash is primarily derived from the incineration of MSW, including household, commercial, and industrial waste, and can reduce waste volume by up to 90%, converting organic content into heat and ash. Careful management of waste incineration ash is essential due to the potential for heavy metal leaching and other contaminants. Proper handling and processing are crucial to prevent environmental pollution. Thus, efficient management or conversion to other applications is vital to reduce its environmental impact [22].

There is potential to substitute cement with MSWIBA due to its chemical composition similarities and specific characteristics [22]. OPC typically contains 69.3% CaO and 15.8% SiO_2, while MSWIBA comprises 31.4% SiO_2 and 29% CaO. Moreover, MSWIBA exhibits higher levels of Al_2O_3, Fe_2O_3, MgO, P_2O_5, SO_3, Cl, K_2O, ZnO, and Pb, indicating a higher concentration of heavy metals compared to OPC [23]. The combined SiO_2 + Al_2O_3 + Fe_2O_3 content in MSWIBA exceeds 50%, meeting ASTM C618 standards for pozzolans, with an SO_3 content of 4.4%, also compliant with ASTM C618 standards [24]. These characteristics suggest that MSWIBA could potentially serve as a viable alternative or supplementary material in concrete production, provided appropriate environmental considerations and processing methods are employed to manage its heavy metal content [25–27].

With these properties, it meets the criteria for use as a cement substitute. Urban waste (municipal solid waste) incineration ash can be utilized in the manufacture of concrete, providing a sustainable alternative to traditional materials. Research has shown that incineration ash from sewage sludge (organic waste) can be used to replace cement in concrete mixes. This promising utilization of waste indicates that, with the right mix proportions, the resulting concrete can achieve adequate strength and flowability, along with sufficient mortar workability for practical applications. Additionally, the leachate characteristics of the resulting concrete are within acceptable levels, further supporting the feasibility of using urban waste as a sustainable alternative in cement production [15].

Experiments were conducted using municipal solid waste bottom ash (MSWIBA) as cement additives at varying replacement levels (0%, 10%, 20%, 30%, 40%, 50%, and 60%), focusing on compressive strength, hydration, microstructure, and heavy metal leaching behaviour in concrete [22]. The research revealed that higher percentages of MSWIBA led to a significant reduction in compressive strength, dropping from 56.81 MPa to 12.65 MPa, attributed to aluminium reactions generating hydrogen in an alkaline environment. Excessive MSWIBA content also resulted in inadequate active substances. Analysis showed an initial increase in ettringite and *calcium silicate hydrate* (*C-S-H*) content (from 0.286% to 0.405% and from 0.635% to 0.893%, respectively), followed by a decline (to 0.275% and 0.500%, respectively). Moreover, compared to conventional concrete, MSWIBA significantly reduced the presence of soluble harmful ions (Cu, Pb, Cr, As, and Hg) by 87.1%, 83.3%, 70.0%, 90.3%, and 89.3%, respectively. Life cycle assessment

(LCA) indicated that using MSWIBA as a supplementary cementitious material (SCM) could decrease environmental impacts (EI) by 7.5% to 44.1% (from 2.58E-11 to 1.56E-11), with reductions observed in global warming potential (GWP) and eutrophication potential (EP), which are major contributors to EI. The trend of decreasing mechanical properties with the increase of cement replacement materials was also observed by *Marieta et al., 2023*. In the experiments conducted, replacing clinker with hydrothermally treated fly ash resulted in a decrease in flexural strength by approximately 60% and compressive strength by approximately 65%. When the fly ash was mixed with calcined and vitrified demolition materials, the flexural strength decreased by approximately 30% and the compressive strength decreased by approximately 50% [28].

4.1.3 Waste Incineration Ash Treatment

MSWI ash cannot be used directly in cement because it contains heavy metals that negatively impact cement properties [29, 30]. The presence of these heavy metals can result in lower compressive strength, reduced durability, and corrosion that damages concrete reinforcing steel [31]. To mitigate these effects, proper processing techniques are necessary to remove or neutralize the harmful elements before the ash can be safely utilized in construction materials. Therefore, processing techniques are necessary to address these issues, one of which is dechlorination to remove hazardous elements from MSWI fly ash [32]. Dechlorination can be achieved through simple washing [33], which involves using water to remove chlorine from the material. Additionally, hydrothermal [34] treatment, which utilizes high temperatures and pressure in an aqueous environment, is also effective in the dechlorination process. This method aims to reduce the chlorine content in the material, making it safer for further use or disposal. More complex methods for removing harmful elements include sintering, melting, and vitrification, or solidification/stabilization [35, 36]. The effectiveness of hydrothermal treatment has been analyzed, with studies concluding that hydrothermal treatment at 200°C for one hour is a useful method for partially dissolving chlorides [37].

Understanding the various methods for heavy metal removal is crucial for selecting the appropriate treatment based on the specific contamination scenario and desired outcomes. Each method has its own set of advantages and limitations, and often, a combination of methods is employed to achieve optimal results. **Figure 4.3** illustrates these various methods, highlighting their approaches to achieving effective heavy metal removal [38].

4.1.4 Recycled Aggregates

In the context of sustainable development, recycling concrete has become a crucial strategy for mitigating environmental impacts and conserving natural resources [39]. Concrete, a fundamental material in the construction

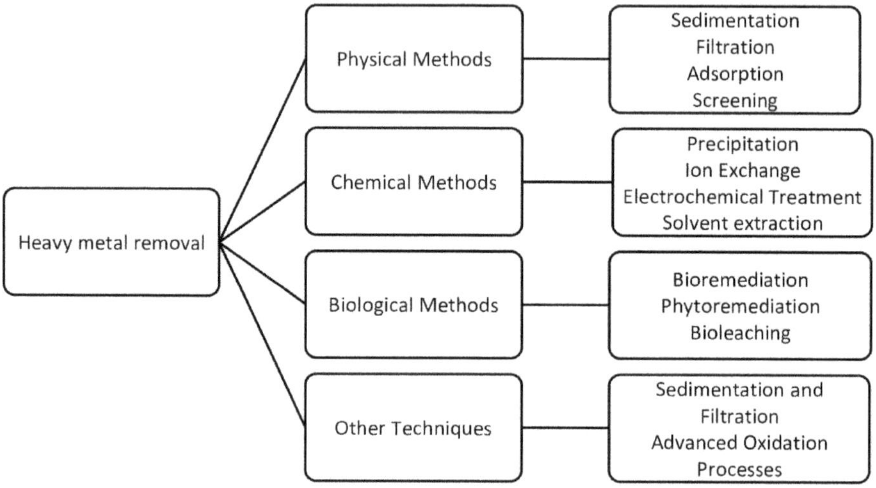

FIGURE 4.3
Various methods for removing heavy metals [37].

industry, often turns into waste or construction debris during the demolition of buildings, reconstruction projects, or renovation activities. This generates pollution problems and increases the demand for landfills, leading to land shortages [40]. The amount of this waste continues to grow, with an estimated 180 million tons produced annually, or 480 kg per person per year. With the rise in construction activities, this trend has been notably observed in Europe [41]. To reclaim the value of these discarded materials, concrete can be refined and processed, particularly the aggregate components, which can then be reused in new concrete mixtures. This recycling process not only reduces landfill waste and associated environmental issues but also promotes the efficient use of resources [42]. Advances in recycling technologies and methodologies have enabled the production of high-quality recycled concrete that meets industry standards [43]. By incorporating recycled concrete into construction projects, the industry can significantly reduce its carbon footprint, enhance resource efficiency, and contribute to a more sustainable and eco-friendly built environment [44].

To achieve high-quality concrete, additional materials such as cement, fly ash (FA), blast furnace slag (BFS), and silica fume (SF) are often required. Research has shown that satisfactory properties in recycled concrete can be attained depending on the quantity and type of supplementary minerals used [42]. Notably, recycled concrete made with 100% recycled coarse aggregate and supplemented with 50% blast furnace slag exhibited slightly higher compressive and split tensile strengths, as well as a lower dynamic modulus of elasticity, compared to recycled aggregate concrete (RAC) using 100% regular Portland cement [42]. **Figure 4.4** shows examples of concrete waste that can be used as concrete aggregate.

FIGURE 4.4
Waste concrete for recycling from laboratory samples and precast column.

Research by *Maier and Durham* [45] also yielded similar results, showing that replacing cement with blast furnace slag (BFS) by up to 50–75% improves the quality of recycled aggregate concrete (RAC). Conversely, substituting cement with fly ash (FA) typically leads to a decline in the mechanical properties of RAC at higher replacement ratios. Analysis by *Lymbachiya* et al. [46] on the effect of partial substitution with fly ash (FA) on recycled aggregate concrete (RAC) with a coarse recycled concrete aggregate (RCA) ratio of 30% shows that replacing 30% of the cement with FA does not adversely affect the mechanical properties of RAC. Additionally, this substitution reduces drying shrinkage, irrespective of the amount of RCA in the concrete.

However, using recycled concrete cannot completely replace cement due to durability concerns, which limit the application of recycled aggregates for structural building purposes. Data indicates that compared to ordinary concrete, recycled aggregate concrete (RAC) experiences an increase in shrinkage of up to 60% [47]. In some cases, using 100% recycled aggregate results in shrinkage of around 70% [48]. According to reports by Olorunsogo and Padayachee, the absorption of recycled air concrete aggregates increased by 47.3%, 43.6%, 38.5%, and 28.8% at curing ages of 3, 7, 28, and 56 days, respectively [49].

4.2 Timber-Steel Hybrid Beams

As part of the green building campaign, one of the efforts being made is to minimize the use of concrete in infrastructure construction. One potential material to replace concrete is wood, which can be used to create timber-steel hybrid structures. This combination is expected to replace the role of concrete steel in infrastructure development. Currently, there is rapid growth in the use of timber for infrastructure building due to its sustainability, excellent seismic behaviour, high ratio of mechanical resistance to material weight, good thermal insulation properties, and ease of fabrication [50, 51]. However, timber has limitations, they are low durability and high hygroscopicity if not properly protected [52]. Therefore, timber construction systems should be made hybrid by combining with different materials' properties to optimize their performance.

Timber exhibits similar properties to concrete in both compression and tension, but at only one-fifth of the weight [53]. The tensile strength of individual cellulose fibres is approximately 8000 N/mm^2, while defect-free wood has a tensile strength of about 100 N/mm^2 [54]. In contrast with other materials, steel is excellent in tension and attractive for high-rise buildings and long-span bridges due to its ductile and predictable properties. It can be easily repaired, prefabricated, reused, and expanded, and it also has good fatigue strength. However, steel has some drawbacks, such as strength loss at high temperatures, rapid heat transmission, susceptibility to corrosion, and a tendency to buckle under compression [54]. By combining the advantages of both materials and addressing their weaknesses, a hybrid wood-steel beam can be produced. This material offers properties such as strength, aesthetics, durability, and lightness, with high compressive strength. It has the potential to evolve and serve as an alternative to concrete.

These hybrid beams typically consist of a primary timber structure reinforced or encased with steel components. In some designs, steel serves as the core material, covered or complemented by timber for aesthetic appeal and additional structural benefits. Hybrid construction can also involve combinations of reinforced concrete, steel, and timber, primarily used for beams, floors, and walls. Timber-concrete composite (TCC) beams are usually formed by coupling a timber beam with reinforced concrete in the transverse (cross-section) or longitudinal direction [55]. This combination allows for the creation of lighter yet stronger structural elements compared to using timber or steel alone. The steel reinforcement significantly increases the load-carrying capacity and structural integrity of the beam, while the use of timber reduces the overall environmental impact. Exposed timber provides a warm, natural look, preferred in architectural designs. Hybrid beams are utilized in residential, commercial, and industrial buildings, bridges, and large-span roof structures, offering both structural support and aesthetic value [56, 57]. However, the production involves precise engineering and manufacturing processes, which can increase overall costs. Despite these challenges, timber-steel hybrid beams offer a promising alternative to traditional building materials, combining strength, sustainability, and visual appeal.

Research on the combination of steel, concrete, and timber is rapidly advancing. Various studies indicate that this combination not only enhances structural strength but also offers aesthetic and sustainability benefits. Hybrid wood-steel technology holds significant potential for high-rise buildings and long-span bridges. Additionally, these structures can reduce overall weight without sacrificing strength, making them ideal for use in earthquake-prone areas. With the development of new technologies and construction methods, the application of hybrid wood-steel beams is expected to become increasingly widespread.

Several studies have investigated the use of timber-concrete composites (TCC) as flooring material. Notably, *Yeoh et al.* [58] examined the mechanical properties of TCC in flooring applications. Other researchers have tested the yield strength of TCC materials, with findings indicating that ultra-high-performance fibre-reinforced concrete (UHPFRC) composite wood exhibits lower deflection compared to traditional wood beams [59]. Additionally, TCC structures have been evaluated under cyclic loading for both bridge and building applications [60]. UHPFRC was used in a wood-concrete hybrid floor with a span of 9 m. The research results show that UHPFRC thin plates can reduce the overall floor weight by 260% and the plate thickness by 20% compared to traditional TCC made with ordinary concrete [61].

The influence of geometry on the mechanical properties of metal-wood construction has been studied [62]. One configuration involved a vertical steel piece placed between two wooden elements (M1), while another used two horizontal steel pieces with wood placed in between (M2). The results showed that M1 experienced a 100% increase in bending strength, while M2 saw an 80% increase in bending stiffness. In terms of load-carrying capacity, M1 exhibited a 100% increase, and M2 showed a 60% increase. In terms of failure modes, all test specimens with horizontal combinations failed due to shearing in the centre of the wood element. The RS specimen material did not reach the yield strength of steel. However, achieving the planned joint stiffness requires complex and expensive joints. **Figure 4.5** presents examples of reinforcement steels bar applied in wood.

Research on the mechanical characteristics of steel-wood composite (STC) structures has been conducted [63]. The materials and geometry tested used glulam wood fastened to steel arches using adhesive. By maintaining a

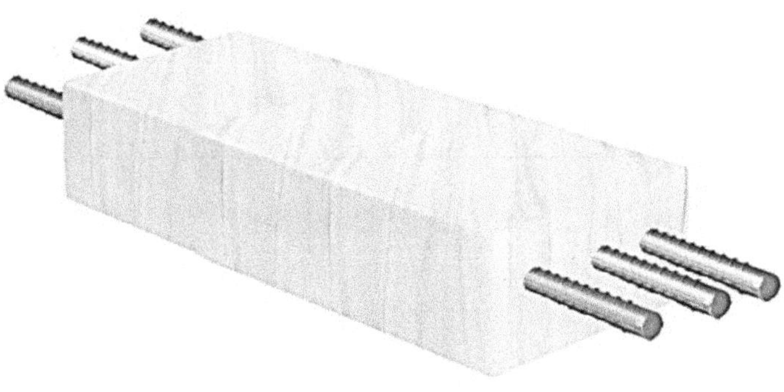

FIGURE 4.5
Example of wood–steel combination material (Illustration: canva, 2024, https://www.canva.com).

steel-to-wood area ratio of 10%, the final load-bearing capacity increased by approximately 28.44%. Finite element analysis was used to examine the influence of steel strength, wood strength, and the steel capacity ratio on ultimate load-bearing capacity. Additionally, this analytical methodology can be used to investigate the properties of STC curvature mechanisms, offering a valuable and straightforward assessment framework.

4.3 Geopolymer Concrete

One effort to replace cement content in concrete is to utilize industrial byproducts to manufacture geopolymer concrete. The composition of geopolymer concrete includes industrial waste materials, such as smelting furnace slag, which contains aluminosilicates. As substitutes for cement, other materials rich in alumina and silica, like fly ash, bottom ash, and rice husk ash, are used. Geopolymer concrete can reduce corrosion current density by 10% compared to ordinary Portland cement (OPC) (*Olivia*, 2011) [64]. The nano-sized porosity of geopolymer concrete withstands water and ion diffusion, and the silicate membranes formed in the paste create a strong protective layer on the reinforcement [65].

Geopolymers are manufactured through the geopolymerization reaction, which involves changing the aluminosilicate raw material into a network consisting of $[-Si-O-Al-O-]_n$ bonds [66]. This geosynthesis reaction is the chemically integrated synthesis of minerals. The geopolymerization reaction forms a binder that can harden to create a geopolymer material. Geopolymers can be made by utilizing different aluminosilicate sources found in nature, such as red mud, blast furnace slag, kaolinite, and rice husk ash. The reaction can occur at room temperature, making it an energy-efficient and cleaner process. Si and Al atoms present in fly ash can be dissolved by hydroxide ions, allowing precursor ions to convert into monomers. These monomers then undergo polycondensation to form polymeric structures.

4.4 Components in Geopolymer Concrete

- Fly ash (FA)

Fly ash is a residue from coal combustion and serves as an effective alternative to manage its abundance. Rich in silicates and alumina, fly ash reacts with alkaline solutions to produce aluminosilicate gel, which binds aggregates to form geopolymers. The fly ash, silica, and alumina in the source material are activated by a base activator to form the gel known as aluminosilicate [67].

- Palm oil fuel ash (POFA)

  POFA, or Palm oil fuel ash, is a byproduct of the palm oil industry, produced when palm oil residues are burnt for energy. This ash contains high amounts of silica, making it valuable for various construction applications. Composition and properties: POFA is rich in silica, which is crucial for its reactivity in concrete and other bonding applications. It may also contain alumina, iron oxide, calcium oxide, and other elements that contribute to its pozzolanic properties.
- Kaolin

  Kaolin is a fine clay rich in aluminosilicate, often used for ceramics. Common aluminosilicate sources include kaolinite, fly ash, calcined kaolin, and chemically synthesized kaolin. Geopolymer is synthesized via polycondensation at 100°C at room pressure in a basic solution. Kaolin is soft, non-abrasive, electrically insulating, and highly resistant, determining the quality of the geopolymer [68].
- Metakaolin

  Metakaolin is a high-reactivity pozzolan derived from the mineral kaolinite, a type of clay. It is produced by the thermal treatment of kaolin at temperatures ranging from 500°C to 900°C, a process known as calcination. This thermal activation removes the chemically bonded water from the kaolin and alters its crystalline structure, creating a highly reactive amorphous aluminosilicate material [68].
- Dolomite

  Dolomite, a soft rock-forming mineral containing calcium *magnesium carbonate* ($CaMg(CO_3)_2$), can be easily crushed into a fine powder and used to reduce dehydration. Dolomite is known as dolostone in its natural form and as metamorphic rock when of high quality. Limestone often contains some dolomite. **Table 4.1** presents the chemical composition of various geopolymer materials and the elements used to form geopolymer materials, respectively.

TABLE 4.1

The Highest Chemical Composition of Geopolymer Materials [67]

Various Types of Geopolymer-based	Highest Chemical Composition	Sources of Raw Materials
Fly ash	SiO_2 & Al_2O_3	Waste product from coal mining
Metakaolin	SiO_2 & Al2O_3	Natural resources
Kaolin	SiO_2 & Al_2O_3	Natural resources
POFA	SiO_2	Waste product from palm oil industry
Dolomites	CaO, MgO & SiO_2	Natural resources

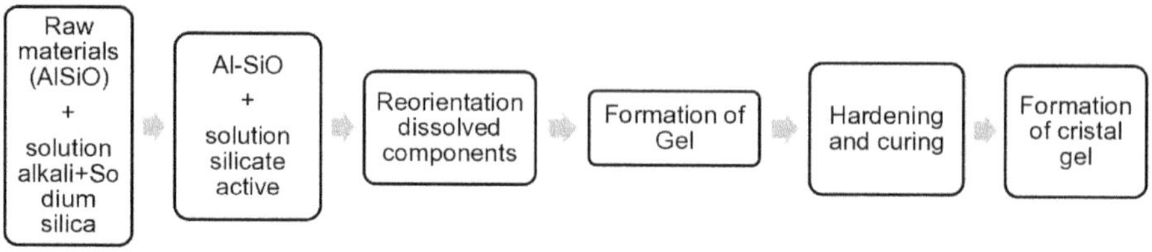

FIGURE 4.6
Representative of the process of forming geopolymer concrete.

The representative process of forming geopolymer concrete is illustrated in **Figure 4.6**. This figure illustrates the key steps involved in the production of geopolymer concrete, a sustainable alternative to traditional Portland cement concrete. The process begins with the selection and preparation of raw materials, typically industrial by-products like fly ash or slag. These materials are then mixed with an alkaline activating solution, such as sodium hydroxide or potassium hydroxide, to initiate the geopolymerization reaction, forming geopolymer concrete.

The main components of geopolymer concrete are presented in **Figure 4.7**. The primary ingredient is an aluminosilicate source, typically obtained from industrial by-products such as fly ash, metakaolin, or slag, which provide the necessary silica and alumina content for the geopolymerization process.

Comparing ordinary concrete (PC) and geopolymer concrete (GPC) across two strength classes (normal 37.5 MPa and high 60 MPa) reveals that GPC can be produced with less water than PC concrete. Consequently, GPC

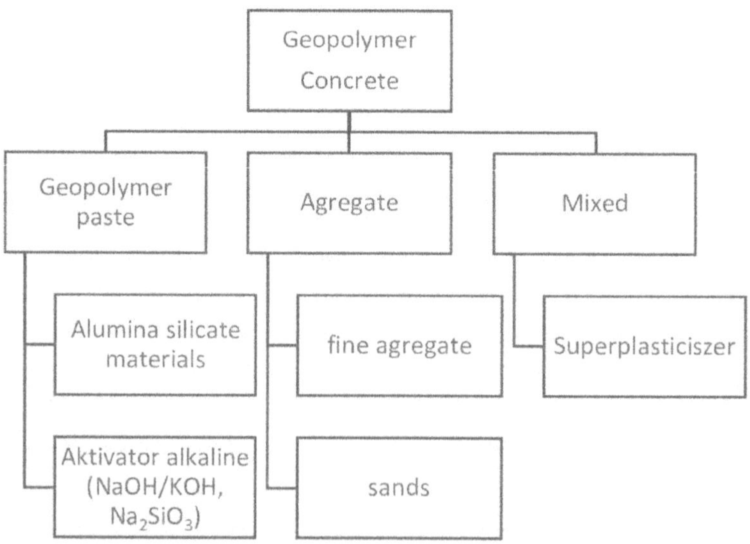

FIGURE 4.7
Composition of geopolymer concrete.

exhibits a faster hardening time and a more rapid increase in compressive strength. By the 28th day, GPC achieves 55–66% of its maximum strength, whereas PC concrete only reaches 18–28% of its maximum strength within the same timeframe [69]. The increase in compressive strength in concrete is influenced by several factors: the binder content, which refers to the amount of fly ash used; the sodium silicate to sodium hydroxide (SS/SH) ratio, which is the proportion of sodium silicate to sodium hydroxide in the mixture; and the alkaline solution to binder ratio, which is the ratio of the alkaline solution to the binder. The use of geopolymer concrete can also reduce the corrosion current density by 10% [64]. The same experiment showed that the geopolymer concrete cracked after nine days while the OPC concrete was cracked after 2.5 days during the electrochemical experiment. Additionally, geopolymer concrete increases flexural strength by up to 100% and reduces the possibility of cracks by 24% [70]. Other benefits include high strength, good creep resistance, low shrinkage coefficient, sulphate resistance, and corrosion resistance [71]. However, geopolymer concrete has weaknesses compared to OPC concrete, particularly in terms of price and curing methods, as it requires high curing temperatures to achieve high strength properties. The steel–cement bond strength in geopolymer concrete is also better. Geopolymer concrete exhibits superior workability, which is typically attributed to the liquid-to-solid ratio in its mixtures, similar to those found in natural pozzolan-based alkali-activated concretes [72].

References

1. Ma, F., Sha, A., Yang, P., & Huang, Y. (2016). The greenhouse gas emission from Portland cement concrete pavement construction in China. *International Journal of Environmental Research and Public Health, 13,* 632.
2. Nielsen, C. V. (2008, May 20–22). *Carbon footprint of concrete buildings seen in the life cycle perspective.* Proceedings of the NRMCA 2008 Concrete Technology Forum: Focus on Sustainable Development, Denver, CO, pp. 1–14.
3. Suhendro, B. (2014). Toward green concrete for a better sustainable environment. *Procedia Engineering, 95,* 305–320.
4. Miller, S. A., Horvath, A., & Monteiro, P. J. M. (2016). Readily implementable techniques can cut annual CO_2 emissions from the production of concrete by over 20. *Environmental Research Letters, 11,* Article 074029.
5. Osial, M., Pregowska, A., Wilczewski, S., Urbańska, W., & Giersig, M. (2022). Waste management for green concrete solutions: A concise critical review. *Recycling, 7*(3), 37. https://doi.org/10.3390/recycling7030037
6. Mohammadhosseini, H., Alyousef, R., & Md. Tahir, M. (2021). Towards sustainable concrete composites through waste valorisation of plastic food trays as low-cost fibrous materials. *Sustainability, 13*(4), 2073. https://doi.org/10.3390/su13042073

7. Awoyera, P. O., Akinmusuru, J. O., & Ndambuki, J. M. (2016). Green concrete production with ceramic wastes and laterite. *Construction and Building Materials, 117,* 29–36.
8. Sizirici, B., Fseha, Y., Cho, C.-S., Yildiz, I., & Byon, Y.-J. (2021). A review of carbon footprint reduction in the construction industry, from design to operation. *Materials, 14*(20), 6094. https://doi.org/10.3390/ma14206094
9. Nielsen, C. V. (2002, June). *Mechanical properties for green concrete.* Featured at the XVIII Symposium on Nordic Concrete Research, Helsingør, Denmark.
10. Uzal, B., Turanli, L., & Mehta, P. K. (2007, September). High-volume natural pozzolan concrete for structural applications. *Materials Journal, 104*(5), 535–538.
11. Maclay, K. (2000, December 27). UC Berkeley retrofits take "green" approach to concrete foundations. *Media Relations.*
12. Berry, M., Cross, D., & Stephens, J. (2009, May 4–7). *Changing the environment: An alternative green concrete produced without Portland cement.* World of Coal Ash (WOCA) Conference, Lexington, KY, USA.
13. Muthaiyan, U. M., & Thirumalai, S. (2017). Studies on the properties of pervious fly ash–cement concrete as a pavement material. *Cogent Engineering, 4.* https://doi.org/10.1080/23311916.2017.1318802
14. Ahmaruzzaman, M. (2010). A review on the utilization of fly ash. *Progress in Energy and Combustion Science, 36,* 327–363. https://doi.org/10.1016/j.pecs.2009.11.003
15. Khankhaje, E., Kim, T., Jang, H., Kim, C.-S., Kim, J., & Rafieizonooz, M. (2023). Properties of pervious concrete incorporating fly ash as partial replacement of cement: A review. *Developments in the Built Environment, 14,* 100130.
16. Mehta, B. R., Sahay, M. K., Malhotra, L. K., Avasthi, D. K., & Soni, R. K. (1996). High energy heavy ion induced changes in the photoluminescence and chemical composition of porous silicon. *Thin Solid Films, 289,* 95–98. https://doi.org/10.1016/S0040-6090(96)08937-7
17. Deepika, S., Lalithanjali, K., Ponmalar, M. R., & Vinushitha, B. (2014). Influence of recycled aggregate based pervious concrete with flyash. *International Journal of ChemTech Research, 7,* 2648–2653.
18. Damtoft, J. S., Glavind, M., & Munch-Petersen, C. (2001, September). *Danish Centre for green concrete.* Proceedings of CANMET / ACI International Conference. San Francisco.
19. Chakraborty, J., & Banerjee, S. (2016). Replacement of cement by fly ash in concrete. *SSRG International Journal of Civil Engineering, 3*(8), 40–42. https://doi.org/10.14445/23488352/IJCE-V3I8P110
20. Liu, H., Luo, G., Wang, L., & Gong, Y. (2019). Strength time-varying and freeze-thaw durability of sustainable pervious concrete pavement material, containing waste fly ash. *Sustainability, 11.* https://doi.org/10.3390/su11010176
21. Spreadbury, C. J., McVay, M., Laux, S. J., & Townsend, T. G. (2021). A field-scale evaluation of municipal solid waste incineration bottom ash as a road base material: considerations for reuse practices. *Resources, Conservation and Recycling, 168,* Article 105264.
22. Magnuson, J. K., Weiksnar, K. D., Patel, A. D., Clavier, K. A., Ferraro, C. C., & Townsend, T. G. (2023). Processing municipal solid waste incineration bottom ash for integration into cement product manufacture. *Resources, Conservation and Recycling, 198,* 107139.

23. Cheng, L., Jin, H., Liu, J., & Xing, F. (2024). A comprehensive assessment of green concrete incorporated with municipal solid waste incineration bottom: Experiments and life cycle assessment (LCA). *Construction and Building Materials, 413,* 134822.
24. ASTM International. (2022). *ASTM C618 standard specification for coal fly ash and raw or calcined natural Pozzolan for use in concrete.* ASTM International.
25. Abe, Y., Shikaku, R., Murakushi, M., Fukushima, M., & Nakai, I. (2021). Did ancient glassware travel the Silk Road? X-ray fluorescence analysis of a Sasanian glass vessel from Okinoshima Island, Japan. *Journal of Archaeological Science: Reports, 40,* Article 103195.
26. Buchner, T., Kiefer, T., Gaggl, W., Zelaya-Lainez, L., & Füssl, J. (2021). Automated morphometrical characterization of material phases of fired clay bricks based on scanning electron microscopy, energy dispersive X-ray spectroscopy, and powder X-ray diffraction. *Construction and Building Materials, 288,* Article 122909.
27. Marieta, C., Martín-Garin, A., Leon, I., & Guerrero, A. (2023). Municipal solid waste incineration fly ash: From waste to cement manufacturing resource. *Materials, 16*(6), 2538. https://doi.org/10.3390/ma16062538
28. Colangelo, F., Cioffi, R., Montagnaro, F., & Santoro, L. (2012). Soluble salt removal from MSWI fly ash and its stabilization for safer disposal and recovery as road basement material. *Waste Management, 32,* 1179–1185.
29. Chen, W. S., Chang, F. C., Shen, Y. H., Tsai, M. S., & Ko, C. H. (2012). Removal of chloride from MSWI fly ash. *Journal of Hazardous Materials, 237,* 116–120.
30. Zhang, S., Chen, Z., Lin, X., Wang, F., & Yan, J. (2020). Kinetics and fusion characteristics of municipal solid waste incineration fly ash during thermal treatment. *Fuel, 279,* 118410.
31. Abiodun, O.-A. O., Oluwaseun, O., Oladayo, O. K., Abayomi, O., George, A. A., Opatola, E., Orah, R. F., Isukuru, E. J., Ede, I. C., Oluwayomi, O. T., et al. (2023). Remediation of heavy metals using biomass-based adsorbents: Adsorption kinetics and isotherm models. *Clean Technologies,* 5(3), 934–960. https://doi.org/10.3390/cleantechnol5030047.
32. Yang, Z., Ji, R., Liu, L., Wang, X., & Zhang, Z. (2018). Recycling of municipal solid waste incineration by-product for cement composites preparation. *Construction and Building Materials, 162,* 794–800.
33. Chen, W. S., Chang, F. C., Shen, Y. H., Tsai, M. S., & Ko, C. H. (2012). Removal of chloride from MSWI fly ash. *Journal of Hazardous Materials, 237,* 116–120.
34. Zhang, S., Chen, Z., Lin, X., Wang, F., & Yan, J. (2020). Kinetics and fusion characteristics of municipal solid waste incineration fly ash during thermal treatment. *Fuel, 279,* 118410.
35. Zhao, K., Hu, Y., Tian, Y., Chen, D., & Feng, Y. (2020). Chlorine removal from MSWI fly ash by thermal treatment: Effects of iron/aluminium additives. *Journal of Environmental Sciences, 88,* 112–121.
36. Marieta, C., Guerrero, A., & Leon, I. (2021). Municipal solid waste incineration fly ash to produce eco-friendly binders for sustainable building construction. *Waste Management, 120,* 114–124.
37. Abdel-Raouf, M. S., & Abdul-Raheim, A. R. M. (2017). Removal of heavy metals from industrial waste water by biomass-based materials: A review. *Journal of Pollution Effects & Control, 5.1.*

38. Radonjanin, V., Malešev, M., Marinković, S., & Al Malty, A. E. S. (2013). Green recycled aggregate concrete. *Construction and Building Materials, 47,* 1503–1511
39. Ahmaruzzaman, M. (2010). A review on the utilization of fly ash. *Progress in Energy and Combustion Science, 36,* 327–363. https://doi.org/10.1016/j.pecs.2009.11.003
40. Marinković, V., Radonjanin, M., Malešev, M., & Ignjatović, I. (2010). Comparative environmental assessment of natural and recycled aggregate concrete. *Waste Management, 30,* 2255–2264.
41. Fisher, C., & Werge, M. (2009). *EU as a recycling society.* ETC./SCP Working Paper 2/2009. Retrieved August 14, 2009, from http://scp.eionet.europa.eu.int
42. Berndt, M. L. (2009). Properties of sustainable concrete containing fly ash, slag, and recycled concrete aggregate. *Construction and Building Materials, 23,* 2606–2613.
43. Xiao, J., Li, W., Fan, Y., & Huang, X. (2012). An overview of study on recycled aggregate concrete in China (1996–2011). *Construction and Building Materials, 31,* 364–383.
44. Maier, P. L., & Durham, S. A. (2012). Beneficial use of recycled materials in concrete mixtures. *Construction and Building Materials, 29,* 428–437.
45. Limbachiya, M., Meddah, M. S., & Ouchagour, Y. (2012). Use of recycled concrete aggregate in fly-ash concrete. *Construction and Building Materials, 27,* 439–449.
46. Hansen, T. C., & Boegh, E. (1985). Elasticity and drying shrinkage of recycled concrete aggregate. *ACI Journal, Proceedings, 82*(5), 648–652.
47. Domingo-Cabo, A., Lázaro, C., López-Gayarre, F., Serrano-López, M. A., Serna, P., & Casta-Tabares, J. O. (2009). Creep and shrinkage of recycled aggregate concrete. *Construction and Building Materials, 23,* 2545–2553.
48. Olorunsogo, F. T., & Padayachee, N. (2002). Performance of recycled aggregate concrete monitored by durability indexes. *Cement and Concrete Research, 32,* 179–185.
49. Arup. (2019). *Rethinking timber buildings - Seven perspectives on the use of timber in building design and construction.* Retrieved September 4, 2021, from Rethinking Timber Buildings: perspectives on the use of timber in building design - Arup
50. Buchanan, A. (2000). Fire performance of timber construction. *Progress in Structural Engineering and Materials, 2,* 278–289.
51. Janssen, H. (2018). A discussion of "Characterization of hygrothermal properties of wood-based products – Impact of moisture content and temperature". *Construction and Building Materials, 185,* 39–43.
52. Asdrubali, F., Ferracuti, B., Lombardi, L., Guattari, C., Evangelisti, L., & Grazieschi, G. (2017). A review of structural, thermo-physical, acoustical, and environmental properties of wooden materials for building applications. *Building and Environment, 114,* 307–332.
53. Mazzucchelli, E. S. (2016). *Sistemi costruttivi in legno - Tecnologie, soluzioni e strategie progettuali verso edifici Zero Energy.* Maggioli Editore. ISBN 978-88-916-1828-3.
54. Augeard, E., Michel, L., & Ferrier, E. (2018). Experimental and analytical study of the mechanical behavior of heterogeneous glulam–concrete beams and panels assembled by a specific treatment of wood. *Construction and Building Materials, 191,* 812–825.
55. Abrahamsen, R. (2017). *Mjøstårnet - Construction of an 81 m tall timber building.* Internationales Holzbau-Forum IHF.

56. Pastori, S., Mazzucchelli, E. S., & Wallhagen, M. (2022). Hybrid timber-based structures: A state of the art review. *Construction and Building Materials, 359,* 129505.
57. Yeoh, D., Fragiacomo, M., De Franceschi, M., & Heng Boon, K. (2011). State of the art on timber-concrete composite structures: Literature review. *Journal of Structural Engineering, 137*(10), 1085–1095.
58. Kong, K., Ferrier, E., Michel, L., & Agbossou, A. (2015). Experimental and analytical study of the mechanical behavior of heterogeneous glulam–UHPFRC beams assembled by bonding: Short- and long-term investigation. *Construction and Building Materials, 100,* 136–148.
59. Balogh, J., Miller, N., Fragiacomo, M., & Gutkowski, R. (2010, June 20–24). *Time-dependent behaviour of composite wood-concrete bridges made from salvaged utility poles.* Proceedings of the 11th World Conference on Timber Engineering, Trentino, Italy.
60. Naud, N., Sorelli, L., Salenikovich, A., & Cuerrier-Auclair, S. (2019). Fostering GLULAM-UHPFRC composite structures for multi-storey buildings. *Engineering Structures, 188*(1), 406–417.
61. Haase, P., Aurand, S., Boretzki, J., Albiez, M., Sandhaas, C., Ummenhofer, T., & Dietsch, P. (2024). Bending behavior of hybrid timber–steel beams. *Materials, 17,* 1164. https://doi.org/10.3390/ma17051164.
62. Zhao, S., Guo, F., Zhao, J., Yang, S., He, F., Liu, H., & Chen, Z. (2023). Investigation of the in-plane mechanical behavior of timber and steel–timber composite arches. *Structures, 57,* 105267.
63. Gourley, J. T., & Johnson, G. B. (2005). Developments in geopolymer precast concrete. *Journal of World Congress Geopolymer,* 139–143.
64. Sobhan, K., Martinez, F. J., & Reddy, D. V. (2021, April 21). Corrosion resistance of fiber-reinforced geopolymer structural concrete in a simulated marine environment. *Canadian Journal of Civil Engineering. 49*(3), 310–317
65. Sitaram, S. (2019). *Advanced geopolymerization technology.* Geopolymers and Other Geosynthetics. http://dx.doi.org/10.5772/intechopen.87250
66. Abdullah, M. M. A., Kamarudin, M., Khairul Nizar, I., & Rafiza, Z. (2011). Review on fly ash-based geopolymer concrete without Portland cement. *Engineering Technology Research,* 3(1), 1–4.
67. Zain, H., & Abdullah, M. M. A. (2017, February). *Review on various types of geopolymer materials with the environmental impact assessment.* MATEC Web of Conferences.
68. Kwasny, J., Soutsos, M., McIntosh, J., & Cleland, D. (2019). Comparison of the effect of mix proportion parameters on behaviour of geopolymer and Portland cement mortars. *Construction and Building Materials, 187,* 635–651.
69. Olivia, M., & Nikraz, H. R. (2011). Corrosion performance of embedded steel in fly ash geopolymer concrete by impressed voltage method. In *Incorporating sustainable practice in mechanics of structures and materials. 1,* 781–786.
70. Zerfu, K., & Ekaputri, J. J. (2016). Review on alkali-activated fly ash-based geopolymer concrete. *Materials Science Forum, 841,* 162–169.
71. Ibrahim, M., Johari, M., Rahman, M. K., & Maslehuddin, M. (2017). Effect of alkaline activators and binder content on the properties of natural pozzolan-based alkali-activated concrete. *Construction and Building Materials, 147,* 648–660.

72. Ibrahim, M., Johari, M., Rahman, M.K., & Maslehuddin, M. (2017). Effect of alkaline activators and binder content on the properties of natural pozzolan-based alkali-activated concrete. *Construction and Building Materials, 147,* 648–660.Top of FormBottom of Form

5

Bio-Based Adhesives

5.1 Introduction

Adhesives function as binders between two objects of different properties to form a unified entity with new characteristics. They have a wide range of applications in both wet and dry conditions. The primary function of adhesives is to provide bond strength, bear loads, and join different materials without requiring mechanical fasteners such as nails or screws. Adhesives have extensive applications in various industries, such as the automotive industry for assembling car components, the construction industry for installing tiles and window glass, the electronics industry for component installation and sealing, and the packaging industry for sealing boxes and labelling products [1–3].

Adhesion theory encompasses several mechanisms, including diffusion theory, which states that objects adhere due to molecular interdiffusion forces between the adhesive surface and the substrate. Meanwhile, adsorption theory posits that adhesion occurs due to adhesive forces between the adhesive molecules and the substrate surface [4]. Chemical interactions, such as van der Waals forces, hydrogen bonds, or chemical bonds, also play a significant role in adhesion. The bonding of an object can also result from mechanical interlocking, where the bond forms due to the physical shape of the adhesive locking with the object's surface. An object may also exhibit electrostatic forces between charged particles on the adhesive surface and the substrate.

In nature, both plants and animals exhibit fascinating methods of adhesion. Plants such as *Hibiscus tiliaceus* use aerial roots to attach to nearby surfaces for support, while *Piper betle* adheres to vertical surfaces using its leaf structure, and *Nelumbo spp.* (lotus) has waxy leaves that allow it to float on water and cling to other leaves. Animals also have impressive adhesion adaptations; geckos possess toe pads with specialized scales enabling them to cling to vertical or upside-down surfaces, spiders use fine hairs on their feet to stick strongly to surfaces and catch prey with their webs, and barnacles attach firmly to hard surfaces like rocks or ship hulls with their feet, resisting ocean waves and acquiring food [5].

DOI: 10.1201/9781003617594-5

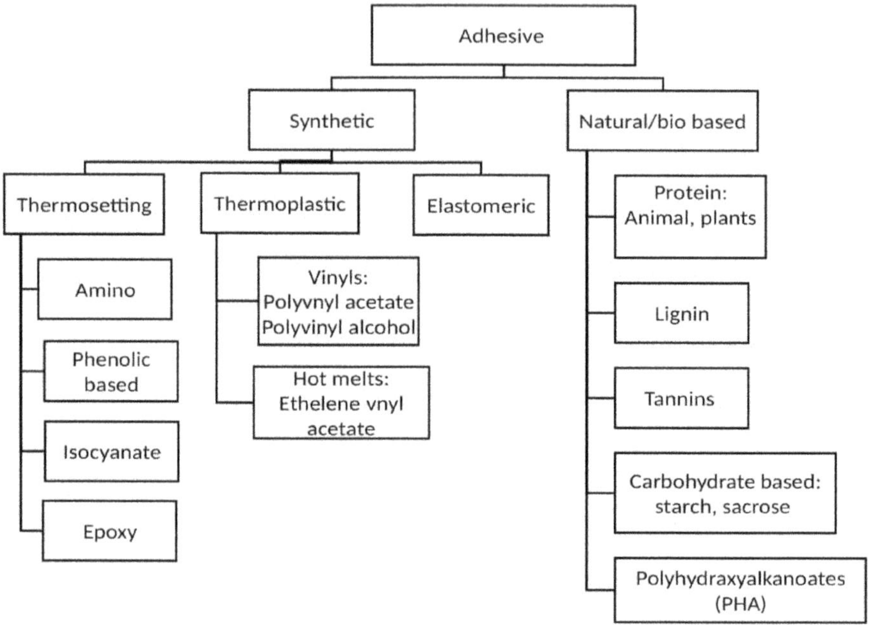

FIGURE 5.1
Classification of adhesives based on raw materials [6].

Adhesives are categorized into natural and synthetic types based on their materials. **Figure 5.1** shows the classification of adhesives by their base materials. Natural adhesives come from plants, animals, or minerals, such as tree resins, milk casein, and animal gelatin. Synthetic adhesives, made in labs, offer specific benefits such as stronger bonds, resistance to extreme temperatures, and better chemical stability. They are durable and can be designed for specific needs but may be less eco-friendly, more expensive, and sometimes contain harmful substances like formaldehyde. Examples of synthetic adhesives include epoxy, super glue, polyurethane, acrylic, silicone, and urea-formaldehyde.

5.2 Natural Adhesive

Natural adhesives are adhesive products made from ingredients found in plants or animals. These natural ingredients are processed to produce

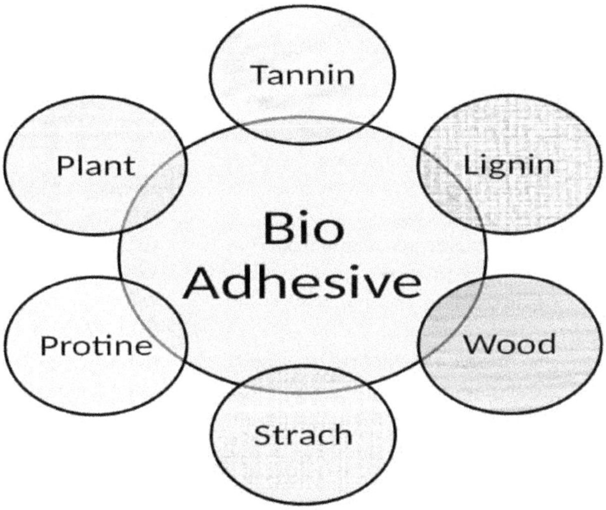

FIGURE 5.2
Materials that can be used as natural adhesives [7, 8].

adhesives that can be used in a variety of applications. Natural-based adhesives are often appreciated for their environmentally friendly properties and safety for users. They can be used in industry and also in everyday life, such as in handicrafts or the manufacture of health and beauty products. Examples of natural-based adhesives include rubber latex, vegetable starch, natural resins, and animal collagen. The advantages of natural-based adhesives include safety, biodegradability, and suitability for applications in various fields [4]. **Figure 5.2** illustrates various materials that can be utilized as natural adhesives.

5.2.1 Natural Adhesive Chemical Compound

An example of a natural adhesive of natural origin is protein, which is obtained from soybeans and has been produced on an industrial scale. This adhesive is divided into two types, namely unmodified (not hydrolysed) and modified (hydrolysed) adhesives, with a classification of high, medium, low, and extra low viscosity. The source of soy protein comes from soybean flakes which have been separated from the fat. Another type of natural adhesive is a material containing starch. Starch is a high molecular weight polymer found in corn, wheat, potatoes, and waxy corn. Starch can be chemically modified for use in surface coatings. **Table 5.1** presents major renewable biopolymers, their sources, and primary industrial applications.

TABLE 5.1

Some Major Renewable Biopolymers, along with Their Sources and Primary Industrial Applications [9]

Biopolymers	Source	Industrial Uses
Cellulose	Trees, plants, plant biomass, plant waste, and byproducts from bio-processing	Textiles, wood manufacturing, and composites
Lignin	Trees, plants, recovered from the pulping processes	Adhesives, coatings, paints, and plastics
Starch	Corn, potato, cassava, wheat, etc.	Adhesives, foams, food, plastics, gums, and pharmaceuticals
Proteins	Soybeans, vegetables, fruits, and animals	Plastics, adhesives, and composites
Oils and waxes	Soybeans, vegetable crops, and specialty crops	Adhesives, resins, coatings, and paints
Chitin	Shellfish, fin fish, and fish waste	Gums, foods, pharmaceutics, and cosmetics
Pectin	Citrus fruits and their waste	Food, gum, emulsifiers, pharmaceutical, and cosmetics.
Latex	Rubber trees or guayule shrubs	Aerospace, medical, plastics, and Adhesives

5.3 Lignin

Lignin is one of the chemical components that make up wood cell walls besides cellulose and hemicellulose. Lignin exists in nature in the form of a polymer composed of phenylpropane units, which have many branches and form a three-dimensional structure. More than two-thirds of the units that make up lignin are linked to each other via ether bonds and the rest is carbon. Lignin is found in cell walls and between cells. The largest lignin concentration is found in the middle lamella layer and decreases towards the secondary layer. The presence of lignin in the cell wall provides cell strength, changes shrinking dimensions, and reduce degradation of cellulose (physical barrier) [10]. Lignin, as a natural phenolic polymer, is abundantly available and is a potential substitute for phenol for the synthesis of phenolic resin adhesives.

In the paper industry, lignin is undesirable so it needs to be removed from the pulp by delignification. The delignification process is used to remove or separate the lignin component from cellulose before the fermentation process is carried out by microbes. Delignification can be carried out using several methods, including using temperature (thermal delignification), acid delignification, and enzymatic or microbial (bio delignification). Thermal delignification can use an autoclave. Many studies have tried to separate high-purity lignin from black liquor and explore its utilization. *Agustini*, [11] in his research, shows how to remove lignin physically (thermally) which is called the delignification process where the biomass is put in an autoclave. Then proceed chemically by adding a 1% H_2SO_4 solution during the process in an autoclave. Meanwhile, biologically, biomass that has been treated in an autoclave is inoculated with microorganisms. The results of the research showed that chemical delignification with 1% H_2SO_4 showed the highest effectiveness of lignin degradation compared to thermally and biologically on *sengon* sawdust and oil palm fronds [12].

5.3.1 Lignocellulose Structure

Lignocellulose is a polysaccharide component that is abundant in nature and consists of three main polymer types: cellulose, hemicellulose, and lignin. Lignocellulose is biomass that comes from plants and is the basic material for plant cell walls [13]. Cellulose, hemicellulose, and lignin form structures called microfibrils, which are then organized into microfibrils. This structure functions to mediate structural stability in plant cell walls, thus playing an important role in plant growth and strength. These three components form complex chemical bonds that provide structural stability to plants. Cellulose is the main component of lignocellulose, which consists of linked beta (1–4) glucose chains. Hydrogen bonds between different layers of these polysaccharides contribute to crystalline cellulose's resistance to degradation. Hemicellulose, the second most abundant component, consists of various 5-carbon and 6-carbon sugars such as arabinose, galactose, glucose, mannose, and xylose. Lignin, which consists of three main phenolic components, namely *p-coumaryl alcohol (H), coniferyl alcohol (G), and sinapyl alcohol (S),* is synthesized through the polymerization of these components. The ratio of lignin components in polymers varies between plants, wood tissues, and different cell wall layers, as can be seen **Figure 5.3**.

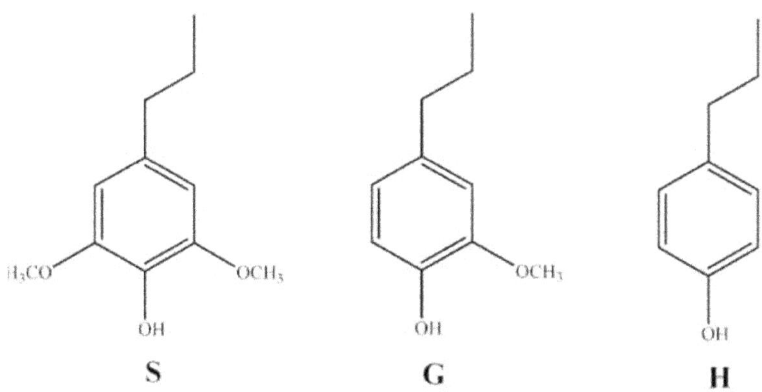

FIGURE 5.3
Structure of the monomers that make up lignin [14, 15].

5.3.2 Lignin Biosynthesis

There are several types of lignin based on how they are isolated, including [16]:

1. Klason Lignin: isolation using concentrated sulphuric acid in the first hydrolysis stage between 68% and 78%, followed by a dilution stage and completed with polysaccharide hydrolysis using a low concentration of acid.
2. Lignin Bjorkman: usually called milled wood lignin, where the wood cell structure is destroyed and the lignin part can be obtained by extracting it by mixing dioxane-water.
3. Cellulolytic enzymatic lignin: polysaccharides are removed using enzymes and the resulting lignin retains its original structure without change.
4. Technical lignin, where lignin is converted into its soluble derivative, intermediate other:
   a. Lignosulphonate: wood is reacted at high temperatures with a solution containing sulphur dioxide and hydrogen sulphite ions.
   b. Kraft lignin and alkaline lignin: the reaction results at a temperature of 170°C with NaOH or a mixture of NaOH and Na_2S.

Agricultural plants contain *cellulose, hemicellulose,* and *lignin* in various agricultural wastes. These important substances can be used for adhesives, as shown in **Table 5.2**.

5.3.3 Ligin Extraction

Lignin can be separated from lignocellulosic components through an extraction process involving physical, chemical, and/or biochemical treatments. Physical treatment generally includes heating in a high-pressure furnace,

TABLE 5.2

Content Cellulose, Hemicellulose, and Lignin in Several Agricultural Wastes and Forest Products [17]

Waste Type	Cellulose (%)	Hemicellulose (%)	Lignin (%)
Broadleaf woody stems Needle-	45–55	24–40	18–25
leafed woody stems	45–50	25–35	25–35
Leaf	15–20	80–85	0
Corncob	45	35	15
Peanut shells	25–30	25–30	30–40
Wheat straw	30	50	15
Bagasse	50	25	25
	41–46	15–33	17–32

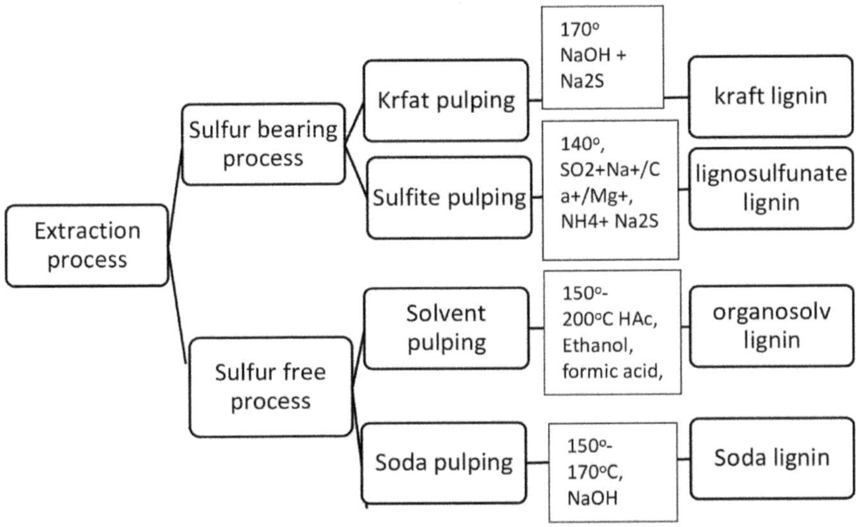

FIGURE 5.4
Method for separating lignin from plants [18] Different extraction processes to separate lignin from lignocellulosic biomass and the corresponding technical lignin [18].

which aims to break the ester and ether bonds in the lignin structure. Chemical and biochemical processes are also used to dissolve and isolate lignin from cellulose and hemicellulose. This extraction produces technical lignin that differs significantly from the natural lignin found in plants. **Figure 5.4** shows the classification of lignin extraction processes into two main categories: sulphur-containing processes and sulphur-free processes. Sulphur-containing processes often involve the use of chemicals such as sulphates or sulphides, while sulphur-free processes do not use such chemicals, which can produce lignin with different properties and wider applications in industry.

5.3.4 Technical Lignin

Technical lignin consists of [6]: Kraft lignin, which is produced on an industrial scale, involves an alkaline chemical process and has a low sulphur content with high purity. Lignosulphonates, also produced industrially, are obtained through an acid process and contain a high sulphur content of low purity. Organosolv lignin, available in industrial and pilot scale, uses a sulphur-free acid process and produces high purity. Lignin Soda, produced on an industrial and pilot scale as well, involves an alkaline process that is sulphur-free and has medium purity. Finally, hydrolytic lignin, also produced on both scales, is obtained through an acid process that can be low sulphur or sulphur-free, and has medium purity. **Figure 5.5** illustrates the current

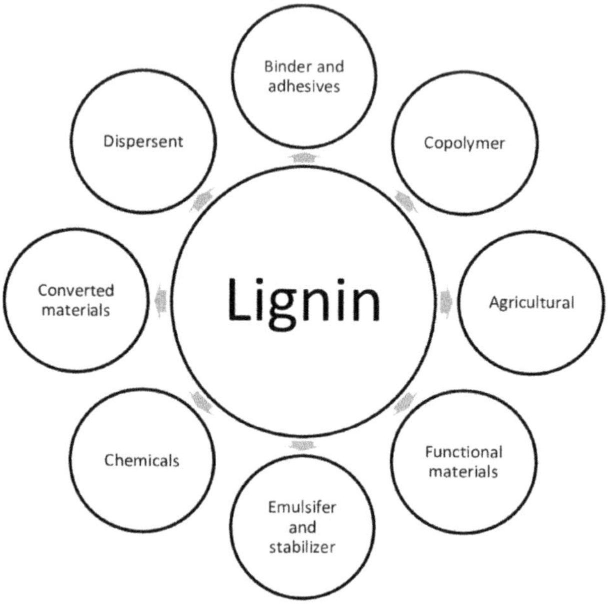

FIGURE 5.5
Current and potential applications of technical lignin [18].

and potential applications of technical lignin for various purposes, highlighting its versatility and growing importance in industrial applications.

5.4 Latex

Latex is a natural polymer derived from the sap of the rubber tree (*Hevea brasiliense*). There are no less than 20,000 types of plants that produce latex, but around 2,500 types of plants are known to contain rubber in their latex [19]. With its super elasticity, high tensile strength and easy shape, rubber makes it a popular material and is widely used in industry [20]. Products ranging from gloves, medical devices, airplanes, vehicle tyres, mattresses, shoe soles, rubber boots, and toys rely on natural rubber as a basic material. Natural rubber is first processed (mixing, creaming, emulsify, vulcanization) so that it can be used to make various kinds of goods that suit the desired properties.

Latex comes from the rubber plant. This plant produces sap which is harvested by tapping, namely by slicing the bark with a knife to obtain latex. The advantage of using natural rubber is that it is biodegradable and the

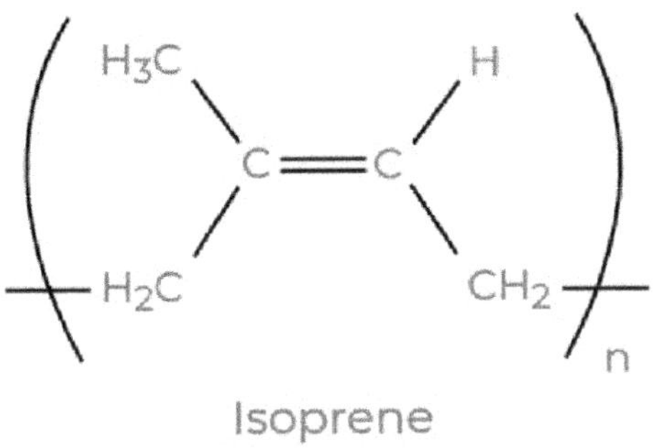

FIGURE 5.6
Image of one of the natural *rubber monomares* (*isoprene*) [20].

trees can be replanted. Natural rubber has isoprene monomer (2-methyl, 1,3-butadiene) as shown in **Figure 5.6.**

5.4.1 Latex Processing

Latex taken from trees consists of about 55% water and about 40% rubber material. Most natural rubber latex comes from one species of rubber tree, *Hevea brasiliense*. The latex composition of different trees varies, but the most common comes from South America. This tree also grows well in Asian plantations. In general, the composition of latex is shown in **Table 5.3.**

The latex taken from the trees is still raw rubber, which does not yet meet the standards for further processing. The initial process carried out to change its composition is by centrifuging concentrated latex (CCL) or by making it into cream through various processes including evaporation, electronation, centrifuging, and creaming. **Table 5.4** displays natural rubber after going through the centrifugal process (CCL). CCL produces 60% rubber, which is an important ingredient in making products such as rubber foam, pillows, mattresses, and elastic bands. Creaming is an environmentally friendly alternative method that uses additives to separate the rubber and serum phases in fresh latex. After the creaming process, the latex composition will change

TABLE 5.3
Composition of Fresh Natural Latex [21]

Constituent	% by Weight of Latex
Rubber particles	30–40
Proteins	2–3
Lipids	0.1–0.5
Sugars	1–2
Others	1.5–3.5
Water	55–65

TABLE 5.4

Latex Content [21]

Materials	Characteristics*	%
Fresh natural latex	TSC	31. 17
	DRC	27.28
Centrifuged concentrated latex	TSC	61.71
	DRC	60.13

TSC: Total solid content
DRC: Dry rubber content
**Tha Chang Industry, Thailand*

TABLE 5.5

Latex Characteristics before and after Concentrated Latex Cream [21]

Properties Fresh	Natural Latex	Concentrated Latex Cream
Total solid content (%)	37.17	46.31
Dry rubber content (%)	27.28	44.61
Alkalinity (%)	0.67	0.14
Volatile fatty acids or VFA	0.042	0.031

as shown in **Table 5.5**. The creaming agents used are sodium carboxymethyl cellulose, poly(vinyl alcohol), methyl cellulose, carboxymethyl cellulose, locust bean gum, tamarind seed powder, casein, and sodium alginate.

5.4.2 Latex Adhesive

Natural rubber latex has long been used as a wood adhesive with quite good bonding qualities. To improve the adhesive quality, natural rubber needs to be modified with other commercial wood adhesives. Natural rubber can also be used as a base polymer for isocyanate adhesives or isocyanate polymer emulsions [22]. The process of modifying the properties of rubber involves several steps, starting with latex coagulation using a coagulant such as formic acid or acetic acid to separate the rubber from water and other materials. The resulting rubber lumps are washed, dried, crushed, and mixed with fillers such as calcium carbonate or kaolin and other additives in a mixing tank. The stirrer ensures that all ingredients are evenly mixed [23]. This mixture is then heated to reduce water content and increase viscosity, as well as aiding the integration of fillers and additives with the latex. The resulting adhesive is quality tested to ensure bond strength, viscosity, pH, and chemical stability meet specifications before being packaged for distribution. Latex adhesives have advantages such as high bond strength, flexibility, resistance to extreme environmental conditions, and are more environmentally friendly than some solvent-based synthetic adhesives. However, production costs may be higher and some types of latex adhesives require additional preservatives to maintain their stability during storage. With these steps, latex is processed into an effective and versatile adhesive, suitable for a variety of industrial and household applications.

Processing latex into rubber products such as car tires, shoe soles, hoses, gaskets, and various other industrial components is done in various ways. The rubber processing process consists of four stages: mixing additives, blending, forming, and vulcanization.

- Admixture: rubber is reinforced with additives and chemicals to increase its tensile strength and characteristics. For example, carbon black filler is added to rubber to increase its tensile strength and protect it from damage from ultraviolet radiation.
- Forming: extrusion, calendaring, moulding or coating, and casting are four common methods for forming rubber goods. Examples of forming processes are casting, spraying, dipping, and electrospinning.
- Vulcanization: process vulcanization consists of various stages: addition of sulphur: Vulcanization generally involves adding sulphur to the rubber. Sulphur forms cross-links between the rubber polymer chains, which changes the molecular structure of the rubber. Heating: after adding sulphur, the rubber is heated to a certain temperature. This heating causes a chemical reaction between the sulphur and the rubber, forming cross-links that make the rubber stronger and more durable. Therefore, the general temperature setting is 60–70°C, with a time range of 4–5 hours (to achieve the required vulcanization degree) [23]. There are several vulcanization methods, namely [23]:
  1. *Yellow sulphur* vulcanization method: in this method, the latex is added with stabilizer, yellow sulphur, zinc oxide, and an accelerator from the vulcanization system, then heated so that latex vulcanization occurs.
  2. *Tyuram vulcanization* method: this method does not use yellow sulphur. Tyuram's vulcanization is carried out with zinc oxide as the vulcanization agent, thiourea and diphenylthiourea as accelerators, and the vulcanization is carried out at a higher temperature (around 90°C). At lower temperatures, the vulcanization process takes place more slowly.
  3. *Organic peroxide* vulcanization method: organic peroxides are used, such as tert-butyl hydroperoxide (t-BHPO) and isopropylbenzene hydrogen peroxide. Generally, more often used together with polyamine compounds (such as tetraethylene pentaamine (TEPA)).

5.5 Cellulose

Cellulose is the main component of plant cell walls, found from higher trees to primitive organisms such as algae, flagellates, and bacteria. Various cellulase-producing microorganisms are: microorganisms from moulds, bacteria, actinomycetes, and acetobacter bacteria [24]. High-purity cellulose can be obtained from a sea animal called tunicin. The structure of cellulose in tunicin is different from cellulose in plant cell walls. Plant cellulose is arranged with the fibre orientation forming certain microfibril angles and forming a helical structure. Meanwhile, tunicin cellulose is arranged randomly or well organized, such as forming a web [13]. Cellulose content can be obtained from various types of algae, namely: *cladophorales* (*cladophora, chaetomorpha, rhizoclonium,* and microdiction) and a few from the class *siphonocladales* (*valonia, dictyosphaeria, siphonocladus,* and *boergesenia*) [25]. Cellulose has properties that make it suitable for use as a base material for adhesives. Cellulose is a straight-chain carbohydrate with glucose as its monomer, linked by hydrogen bonds with the chemical formula $(C_6H_{10}O_5)n$. Cellulose is insoluble in various solvents and is resistant to many chemicals, except strong acids, due to the presence of hydrogen bonds between hydroxyl groups in the cellulose chain. The growing environment contributes to variations in plant cell wall components, including cellulose [13]. Cellulose is the main element in plants in the form of a linear biopolymer of *anhydroglucopyranose molecules in β-1,4 glucosidic* bonds which is abundant in nature [24, 26]. There are four main groups of cellulose sources based on type: wood cellulose, non-wood, marine fauna, and bacterial activity [27]. Each source has variations in cellulose content and purity. Each plant has a different cellulose composition depending on the part of the plant and the type (**Table 5.6**).

TABLE 5.6

Cellulose Content in Agricultural Waste [24, 28]

Lignocellulosic Materials	Cellulose (%)
Hardwood stem	40–55
Softwood stalk	45–50
Kanagan skin	25–30
Corn cobs	45
Paper	85–99
Wheat straw	30
Rice straw	32.1
Garbage disposal	60
Leaf	15–20
Cotton seed hairs	80–95

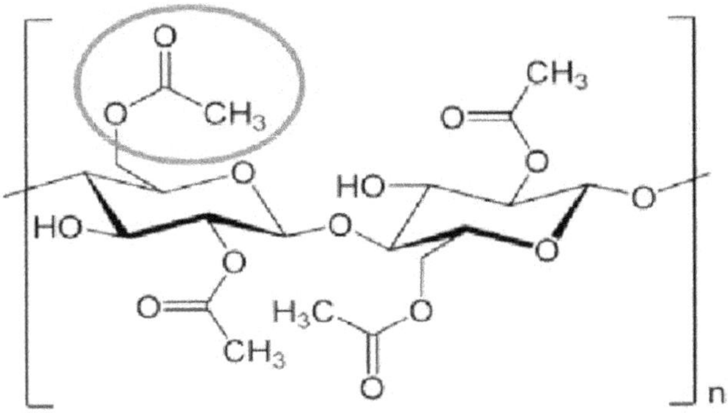

FIGURE 5.7
Cellulose acetate [29].

5.5.1 Cellulose Structure

Cellulose has linear chains through intermolecular and intramolecular hydrogen bonds. The *OH* groups in cellulose determine the physical and chemical properties of cellulose. **Figure 5.7** shows the structure of cellulose acetate. Cellulose has uniform units and bonds, where the basic unit consists of two glucose anhydride units called cellobiose units. Cellulose polymers have an average DP between 300–3000 and an average molecular weight between 50.000–500.000 g/mol. When undergoing the hydrolysis process, the molecular weight of microcrystalline cellulose decreases to between 30.000 and 50.000 g/mol due to chain breaking, so that the DP of cellulose becomes less than 400. The cellulose polymer consists of two main parts: a crystalline part with a regular structure and an amorphous part with an irregular structure. The crystalline structure of cellulose influences the physical and mechanical properties of cellulose fibres, with ordered molecular arrangements punctuated by irregular arrangements every 60 nm, allowing folding of the cellulose chains. Repeated drying and wetting of cellulose will increase cross-linking between microfibrils due to the addition of hydrogen bonds [29].

There are four types of cellulose: pure, natural, commercial (pulp), and laboratory. Pure cellulose is found in cotton, natural cellulose is in wood and non-wood fibres. Commercial cellulose is obtained through pulping of lignocellulosic biomass, while laboratory cellulose is obtained through extraction and reaction with sodium chlorite and alkali. Each type has different characteristics and cellulose content.

5.5.2 Cellulose Extraction

Cellulose extraction can be carried out through several methods, namely mechanical, biological, and enzymatic processes, as well as chemical processes. The following is an explanation of each method:

1. Mechanical process [30–32]:
    - Hi-pressure homogenizer: this process uses high pressure to break down plant cells or biomass into smaller particles, allowing better access to the cellulose fibres.
    - Stone grinder: stone grinders are used to mechanically crush biomass into fine cellulose fibres.
    - Ultrasonication: this method uses ultrasonic waves to break down plant cells and separate the cellulose fibres from the matrix.
2. Biological and enzymatic processes:
    - Biological and enzymatic processes involve the use of microorganisms or enzymes to break down the plant matrix and isolate the cellulose fibres. Certain enzymes can be used to break down chemical bonds between plant cell components and increase the yield of cellulose isolation.
3. Chemical processes [33]:
    - Acid hydrolysis: cellulose is broken down using a strong acid such as sulphuric acid or hydrochloric acid to separate the cellulose fibres from lignin and hemicellulose.
    - Carboxymethylation: this process involves the reaction of cellulose with *sodium hydroxide* and *monochloroacetate* to produce carboxymethyl cellulose, which has specific properties for specific applications such as in the food, pharmaceutical, or chemical industries.
    - Oxidation tempo [34]: oxidation with *2,2,6,6-tetramethylpiperidine-1-oxyl* (TEMPO) is used to change the surface properties of cellulose fibres, increasing their dispersion ability and chemical reactivity.
    - Periodate-chlorite oxidation: this method involves using a mixture of periodate and chlorite to oxidize cellulose side chains, producing products with a variety of functional properties.

5.6 Chitosan

Chitin and chitosan are polysaccharides (long chains of complex carbohydrates) obtained from animals or microbes. Chitoasan is the most common chitin derivative, produced from the chitosan deacetylation process (removal of acetyl groups (*-$COCH_3$*)). When extracted from microbial sources, these biopolymers are usually obtained in complex form, in which chitosan/chitin is covalently linked to glucan chains. Chitin has been developed because the biocompatibility, biodegradability, and biological activity of this polysaccharide make it an attractive biomaterial for various applications. In industrial processing, chitin is extracted through acid treatment to dissolve calcium carbonate, followed by an alkaline solution to dissolve proteins. Additionally, a decolorization step is often added to remove pigments and obtain pure, colourless chitin. The insolubility of chitosan in aqueous solutions limits its use in various applications. To overcome this problem, chemical modification of chitosan was carried out [35].

Chitin has been widely applied in gel form which is useful in the fields of biomedicine, pharmacy, food technology, and cosmetics. Chitosan, which is biodegradable, antibacterial, non-toxic, and helps heal wounds and bones, makes it ideal for composite gels. To adapt properties such as swelling, porosity, mechanical strength, and drug release, chitosan is mixed with other materials such as hydrogels, polymers, or nanoparticles. In gel form, chitin can act as a medium for wound healing, tissue engineering, drug delivery systems, 3D printing, and regenerative medicine.

In the pharmaceutical field, chitin gel functions to improve drug delivery systems by improving bioavailability and allowing longer drug release. Chitosan can also be used in the manufacture of mucoadhesive films, transdermal patches, and buccal drug delivery systems that allow better drug absorption and long-lasting effects [36]. In addition, chitosan composite gel has important applications in food technology as a natural preservative that can extend the shelf life of food products and in cosmetics as a moisturizing agent and binding agent that is safe for the skin. The sustainability and safety of chitosan make it an attractive choice for a variety of industrial applications. In the field of food technology, chitosan is used as a natural preservative that can extend the shelf life of food products. Its antibacterial properties help prevent the growth of pathogenic microorganisms, while its edible nature makes it safe to consume. In cosmetics, chitosan is used as a moisturizer and binding agent that is safe for the skin. Its ability to form a thin film on the surface of the skin helps retain moisture and provides a protective effect.

5.6.1 Chitin Structure

The chemical structure of chitosan consists of several *2-acetamido-2-deoxy-β-d-glucopyranose* units (**Figure 5.8**). This structure is similar to cellulose, with the difference that at the C2 position, chitosan has an acetamide group (–$NHCOCH_3$) instead of the hydroxyl. Although some glucopyranose residues are in the deacetylated form as *2-amino-2-deoxy-β-d-glucopyranose,* chitosan is considered a homopolymer. The degree of *acetylation* (DA) is usually around

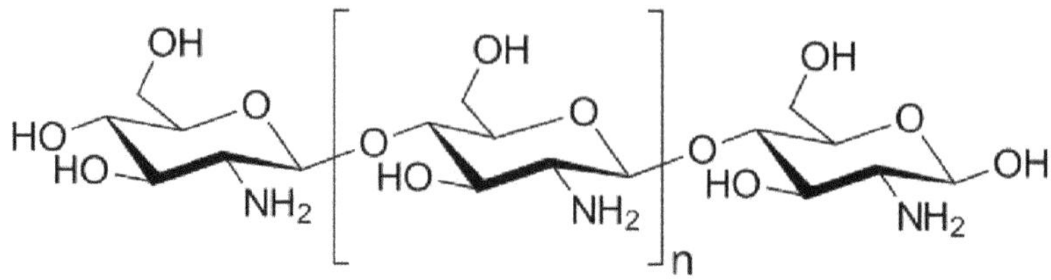

FIGURE 5.8
Chitosan structure

0.90, indicating the presence of some amino groups (5–15%) [37]. When the degree of acetylation of chitosan is low enough that protonation of the amine produces a soluble polysaccharide, chitosan is considered a distinct chemical entity, known as chitosan [38]. Chitin morphology depends on the source; chitin is found in two allomorphs, namely α and β forms. The *α-Chitin isomorph* is by far the most abundant; it is found in the cell walls of fungi and yeast, in the tendons of krill, lobster, and crab, in the shells of shrimp, and in the cuticle of insects. In addition to natural chitin, α-chitin is systematically formed by recrystallization from chitin solutions, in vitro biosynthesis, or enzymatic polymerization due to the high thermodynamic stability of this isomorph [39].

5.6.2 Chitin Extraction

The main source of raw materials for chitin production is the shells of various crustaceans, especially crabs and shrimp. In shellfish, chitin is found as a constituent of a complex network with proteins in which calcium carbonate is deposited to form a rigid shell [40]. Isolation of chitin from shellfish requires the removal of the two main constituents of the shell, protein through deproteination and inorganic calcium carbonate through demineralization, along with small amounts of pigments and lipids that are generally removed during the previous two steps [40]. In some cases, an additional decolorization step is applied to remove residual pigment.

5.6.3 Chemical Extraction

a. Chemical deproteination

This is done heterogeneously using chemicals that also depolymerize the biopolymer. Complete removal of the protein is critical for biomedical applications, as a percentage of the human population is allergic to shellfish, the primary cause of which is the protein component. Chemical methods are the first approach used in deproteination. Various chemicals have been tested as deproteination reagents including *NaOH*, Na_2CO_3, $NaHCO_3$, *KOH*, K_2CO_3, $Ca(OH)_2$, Na_2SO_3, $NaHSO_3$, $CaHSO_3$, Na_3PO_4, and Na_2S [41].

b. Chemical demineralization

Demineralization consists of the removal of minerals, especially calcium carbonate. Demineralization is generally carried out by acid treatment using *HCl*, HNO_3, H_2SO_4, CH_3COOH, *and HCOOH.* Among these acids, the most preferred reagent is dilute hydrochloric acid. Demineralization is easily achieved because it involves the decomposition of calcium carbonate into water-soluble calcium salts with the release of carbon dioxide [42].

5.6.4 Biological Extraction of Chitin

Using proteolytic enzymes to digest proteins or fermentation processes using microorganisms that allow digestion of both proteins and minerals [43].

a. Enzymatic Deproteination: Chitin extraction requires the use of proteases. Proteolytic enzymes originate primarily from plant, microbial, and animal sources. Many proteases such as *alcalase, pepsin, papain, pancreatin, devolvase,* and *trypsin* remove proteins from crustacean shells and minimize deacetylation and depolymerization during chitin isolation [44].
b. Fermentation: the cost of using enzymes can be reduced by deproteinizing them through a fermentation process, which can be achieved by endogenous microorganisms (called auto-fermentation) or by adding specific strains of microorganisms [45]. This can be achieved through single-stage fermentation, two-stage fermentation, co-fermentation, or successive fermentation. Fermentation methods can be divided into two main categories: lactic acid fermentation and non-lactic acid fermentation. *Lactic Acid* Fermentation: Fermentation of crustacean shells can be done with strains of Lactobacillus sp. chosen as the inoculum that produces lactic acid and protease [46]. Non-Lactic Acid Fermentation: In non-lactic acid fermentation, bacteria and fungi are used to ferment *crustacean shells,* for example: *Bacillus sp., Pseudomonas sp., and Aspergillus sp* [47].

References

1. Anghvi, M., Tambare, O., & More, A. (2022). Performance of various fillers in adhesives applications: A review. *Polymer Bulletin, 79*(1), 1–63.

2. Lamberti, M., Maurel-Pantel, A., & Lebon, F. (2023). Mechanical performance of adhesive connections in structural applications. *Materials, 16*(22), 7066. https://doi.org/10.3390/ma1622706
3. He, X. (2011). A review of finite element analysis of adhesively bonded joints. *International Journal of Adhesion and Adhesives, 31*(4), 248–264.
4. De Buyl, F. (2001). Silicone sealants and structural adhesives. *International Journal of Adhesion and Adhesives, 21*(5), 411–422.
5. Sugiman. (2022). *Ilmu dan Teknologi Adhesi.* Depublish.
6. Ebnesajjad, S. (2009). Chapter 4 - classification of adhesives and compounds. In *Adhesives technology handbook* (2nd ed., pp. 47–62). William Andrew Publishing.
7. Inamuddin, Boddula, R., Ahamed, M. I., & Asiri, A. M. (Eds.). (2020). *Green adhesives: Preparation, properties, and applications.* John Wiley & Sons, Incorporated.
8. Magalhães, S., Alves, L., Medronho, B., Fonseca, A. C., Romano, A., Coelho, J. F. J., & Norgren, M. (2019). Brief overview on bio-based adhesives and sealants. *Polymers, 11*(10), 1685. https://doi.org/10.3390/polym11101685
9. Imam, S. H., Bilbao-Sainz, C., Chiou, B. S., Glenn, G. M., & Orts, W. J. (2012). Biobased adhesives, gums, emulsions, and binders: current trends and future prospects. *Journal of Adhesion Science and Technology, 27*(18–19), 1972–1997. https://doi.org/10.1080/01694243.2012.696892
10. Theadioke Center. (2010, May 17). *Lignin.* https://theadiokecenter.wordpress.com/2010/05/17/lignin/
11. Agustini, L., & Efiyanti, L. (2015). Pengaruh Perlakuan Delignifikasi Terhadap Hidrolisis Selulosa Dan Produksi Etanol Dari Limbah Berlignoselulosa. *Jurnal Penelitian Hasil Hutan, 33*(1), 69–80. http://dx.doi.org/10.20886/jphh.v33i1.640.69-80
12. Rahhutami, R., Handini, A. S., & Lestari, I. (2020). Pengaruh Delignifikasi Termal terhadap Substansi Dinding Sel pada Limbah Bunga Jantan Kelapa Sawit Pasca Anthesis. *Jurnal agro industri Perkebunan, 8*(2), 1377. https://doi.org/10.25181/jaip.v8i2.1377
13. Fatriasari, W., Masruchin, N., & Hermiati, E. (2019). *Selulosa: Karakteristik dan Pemanfaatannya.* LIPI Press.
14. Wu, W., Li, P., Huang, L., Wei, Y., Li, J., Zhang, L., & Jin, Y. (2023). The role of lignin structure on cellulase adsorption and enzymatic hydrolysis. *Biomass, 3*(1), 96–107. https://doi.org/10.3390/biomass3010007
15. Mandlekar, N., Cayla, A., Rault, F., Giraud, S., Salaün, F., Malucelli, G., & Guan, J.-P. (2018). *An overview on the use of lignin and its derivatives in fire retardant polymer systems.* InTech. https://doi.org/10.5772/intechopen.72963
16. Nofriadi, E. (2009). *Keragaman Nilai Lignin Terlarut Asam (Acid Soluble Lignin) Dalam Kayu Reaksi Pinus Merkusii Jungh Et De Vriese Dan Gnetum Gnemon Linn* (Unpublished undergraduate thesis). Institut Pertanian Bogor, Bogor, Indonesia.
17. Hermiati, E., Mangunwidjaja, D., Sunarti, T. C., Suparno, O., & Prasetya, B. (2010). Pemanfaatan Biomassa Lignoselulosa Ampas Tebu untuk Produksi Bioetanol. *Jurnal Litbang Pertanian, 29*(4). https://media.neliti.com/media/publications/178805-ID-none.pdf
18. Mandlekar, N., Cayla, A., Rault, F., Giraud, S., Salaün, F., Malucelli, G., & Guan, J.-P. (2018). *An overview on the use of lignin and its derivatives in fire retardant polymer systems.* InTech. https://doi.org/10.5772/intechopen.72963

19. Arias, M., & van Dijk, P. J. (2019). What is natural rubber and why are we searching for new sources?. *Frontiers for Young Minds, 7,* 100. https://doi.org/10.3389/frym.2019.00100
20. Andrade, K. L., Ramlow, H., Floriano, J. F., et al. (2023). Latex and natural rubber: processing techniques for biomedical applications. *Brazilian Journal of Chemical Engineering, 40*(4), 913–927. https://doi.org/10.1007/s43153-023-00317-y
21. Suksup, R., Imkaew, C., & Smitthipong, W. (2017). *IOP Conference series: Materials science and engineering.* 4th International Conference on Mechanical, Materials and Manufacturing (ICMMM 2017), Vol. 272, pp. 25–27, Atlanta.
22. Hermiati, E., et al. (2019). Optimization of application of natural rubber based API adhesive for the production of laminated wood. *IOP Conference Series: Earth and Environmental Science, 374,* 012007. https://doi.org/10.1088/1755-1315/374/1/012007
23. K., D. (n.d.). *Principles of latex formulation design and vulcanization process (III).* NEWSLETTER, Rubber & Latex Industry Knowledge, International Trade Managers - Company Specialized in Producing & Exporting PVC.
24. Anindyawati, T. (2010). Potensi selulase dalam mendegradasi lignoselulosa limbah pertanian untuk pupuk organik. *Berita Selulosa, 45*(2), 70–77.
25. Mihranyan, A., Edsman, K., & Strømme, M. (2007). Rheological properties of cellulose hydrogels prepared from Cladophora cellulose powder. *Food Hydrocolloids, 21*(2), 267–272.
26. Dashtban, M., Schraft, H., & Qin, W. (2009). Fungal bioconversion of lignocellulosic residue: Opportunities & perspectives. *International Journal of Biological Sciences, 5*(8), 578–595.
27. Nechyporchuk, O., Belgacem, M. N., & Bras, J. (2016). Production of cellulose nanofibrils: A review of recent advances. *Industrial Crops and Products, 93,* 2–25.
28. Howard, R. L., Abotsi, E., van Rensburg, E. L., & Howard, S. (2003). Lignocellulose biotechnology: Issues of bioconversion and enzyme production. *African Journal of Biotechnology, 2*(12), 602–619.
29. Tayeb, A. H., Amini, E., Ghasemi, S., & Tajvidi, M. (2018). Cellulose nanomaterials—Binding properties and applications: A review. *Molecules, 23*(10), 2684. https://doi.org/10.3390/molecules23102684
30. Turbak, A. F., Snyder, F. W., & Sandberg, K. R. (1983). Microfibrillated cellulose, a new cellulose product: Properties, uses and commercial potential. *Journal of Applied Polymer Science: Applied Polymer Symposium, 37,* 815–827.
31. Nechyporchuk, O., Belgacem, M. N., & Bras, J. (2016). Production of cellulose nanofibrils: A review of recent advances. *Industrial Crops and Products, 93,* 2–25.
32. Iwamoto, S., Nakagaito, A. N., & Yano, H. (2007). Nano-fibrillation of pulp fibers for the processing of transparent nanocomposites. *Applied Physics A, 89*(2), 461–466.
33. Pääkkö, M., Ankerfors, M., Kosonen, H., Nykäenen, A., Ahola, S., Oesterberg, M., Ruokolainen, J., & Lindström, T. (2007). Enzymatic hydrolysis combined with mechanical shearing and high-pressure homogenization for nanoscale cellulose brils and strong gels. *Biomacromolecules, 8*(6), 1934–1941.
34. Saito, T., Nishiyama, Y., Putaux, J.-L., Vignon, M., & Isogai, A. (2006). Homogenous suspensions of individualized microfibrils from TEMPO-catalyzed oxidation of native cellulose. *Biomacromolecules, 7*(6), 1687–1691.

35. Farinha, I., & Freitas, F. (2020). Chemically modified chitin, chitosan, and chitinous polymers as biomaterials. In S. Gopi, S. Thomas, & A. Pius (Eds.), *Handbook of chitin and chitosan* (pp. 43–69). Elsevier.
36. Bonde, S., Chandarana, C., Prajapati, P., & Vashi, V. (2024). A comprehensive review on recent progress in chitosan composite gels for biomedical uses. *International Journal of Biological Macromolecules, 272*, 132723. https://doi.org/10.1016/j.ijbiomac.2024.132723
37. Franca, E., Lins, R., Freitas, L., & Straatsma, T. (2008). Characterization of chitin and chitosan molecular structure in aqueous solution. *Journal of Chemical Theory and Computation, 4*(12), 2141–2149.
38. Hajji, S., Younes, I., Ghorbel-Bellaaj, O., Hajji, R., Rinaudo, M., Nasri, M., et al. (2014). Structural differences between chitin and chitosan extracted from three different marine sources. *International Journal of Biological Macromolecules, 65*, 298–306.
39. Rudall, K. M., & Kenchington, W. (1973). The chitin system. *Biological Reviews, 40*(4), 597–636. https://doi.org/10.1111/j.1469-185X.1973.tb01570.x
40. Horst, M. N., Walker, A. N., & Klar, E. (1993). The pathway of crustacean chitin synthesis. In M. N. Horst & J. A. Freeman (Eds.), *The crustacean integument: Morphology and biochemistry* (pp. 113–149). CRC Press.
41. Younes, I., & Rinaudo, M. (2015). Chitin and chitosan preparation from marine sources. Structure, properties and applications. *Marine Drugs, 13*(3), 1133–1174. https://doi.org/10.3390/md13031133
42. No, H. K., & Hur, E. Y. (1998). Control of foam formation by antifoam during demineralization of crustacean shell in preparation of chitin. *Journal of Agricultural and Food Chemistry, 46*(9), 3844–3846. https://doi.org/10.1021/jf9802676
43. Khanafari, A., Marandi, R., & Sanatei, S. (2008). Recovery of chitin and chitosan from shrimp waste by chemical and microbial methods. *Iranian Journal of Environmental Health Science & Engineering, 5*(1), 1–24.
44. Rao, M. B., Tanksale, A. M., Ghatge, M. S., & Deshpande, V. V. (1998). Molecular and biotechnological aspects of microbial proteases. *Microbiology and Molecular Biology Reviews, 62*(3), 597–635.
45. Arbia, W., Arbia, L., Adour, L., & Amrane, A. (2013). Chitin extraction from crustacean shells using biological methods—A review. *Food Technology and Biotechnology, 51*(1), 12–25.
46. Prameela, K., Murali Mohan, C., Smitha, P. V., & Hemablatha, K. P. J. (2010). Bioremediation of shrimp biowaste by using natural probiotic for chitin and carotenoid production: An alternative method to hazardous chemical method. *International Journal of Applied Biology and Pharmaceutical Technology, 1*(3), 903–910.
47. Mahmoud, N. S., Ghaly, A. E., & Arab, F. (2007). Unconventional approach for demineralization of deproteinized crustacean shells for chitin production. *American Journal of Biochemistry and Biotechnology, 3*(1), 1–9. https://doi.org/10.3844/ajbbsp.2007.1.9

6

Utilization of Rice Husk Ash: Sustainable Solutions for the Environment

6.1 Rice Husk Ash (RHA)

In agriculture, rice husks are the hard protective layer that protects rice grains and is removed during the milling process. Rice husks are a waste material widely available in all rice-producing countries and contain approximately 30 to 50% organic carbon by weight. Rice husk ash (RHA) is a byproduct of burning rice husks. During the burning process, most of the volatile components of rice husks are gradually destroyed, leaving only silicates as the main residue. The rice husk content, burning temperature, and burning duration all influence the properties of the ash. For example, for every 100 kg of husk burnt in a boiler, about 25 kg of RHA is produced. Some regions utilize rice husks as fuel for the rice milling process, while others burn them in the fields as a local heat and fuel source. However, burning rice husks in these situations is far from perfect, and this incomplete combustion contributes to air pollution.

Before being utilized, rice husks are first converted into ash, from which the silica content is subsequently extracted [1, 2]. Furthermore, silica from rice husk ash can be used for various applications, as shown in **Figure 6.1**.

Rice husks constitute almost 20% of the total weight of rice. Rice husk is a byproduct of rice farming whose structure consists of lignin (25–30%), cellulose (50%), moisture (10–15%), and silica (15–20%) [4]. Rice husks have a heating value of about half that of coal, and assuming the husks contain around 8t–10% water content and no bran, the calorific value is calculated to be around 16.7 MJ/kg [5]. Rice husks have a calorific value that is approximately half that of coal. By burning rice husks under controlled conditions and over a long period of time, volatile organic materials, consisting mainly of cellulose and lignin, are removed, and the remaining ash consists mainly of amorphous silica with a cellular (microporous) structure. The specific surface area of RHA evaluated by the *Brunauer–Emmett–Teller (BET) nitrogen adsorption technique* can vary from 20 to as high as 270 m^2/g, while the specific surface area of silica fume, for example, can range from 18 to 23 m^2/g. [6].

DOI: 10.1201/9781003617594-6

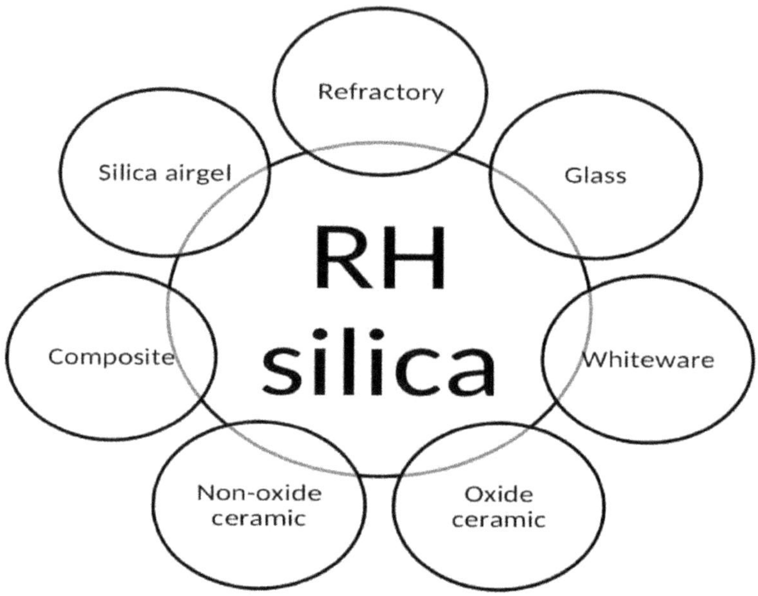

FIGURE 6.1
Utilization of rice husk ash in the form of silica for various applications [3].

Figure 6.2 illustrates the transformation process of rice husks into silica. The process begins with burning rice husks to produce husk ash, which is then extracted to extract the silica content. Furthermore, the silica obtained through this process can be used for various industrial applications.

Paddy husk ash can be used in various products across the agriculture and manufacturing industries. This material can be utilized effectively as a fertilizer, widely applied to enhance plant growth. In the concrete industry, rice husk ash is used to make the material more fire-resistant, providing an excellent alternative to other types of ash. As an industrial product, paddy husk ash is available in various forms, offering multiple utilities for different applications. It contains essential chemical elements required for producing fertilizers or basic industrial materials, particularly silica and ceramic alloys (**Table 6.1**).

6.1.1 Extraction of Rice Husk Ash into Silica

Silica is a naturally occurring mineral that forms from the decomposition of organic materials of fossil origin by biotic, abiotic, or geological agents such as bacteria, fungi, or animals (known as metathesis). Most silica is in crystalline form, where the type of crystal and its properties depend on the surrounding temperature and pressure. Silica is widely used in industry for insulation, glass reinforcement, and other construction materials such as

FIGURE 6.2
Shape husk paddy become silica, (a) husk paddy, (b) ash husk, ash husk powder, silica.

TABLE 6.1
Composition of Rice Husk Ash [7–9]

Chemical Constituent	Rice Husk Ash Weight (%)
SiO_2	82.14–96.34
Al_2O_3	1.34–0.41
Fe_2O_3	1.27–0.20
CaO	1.21–0.41
MgO	0.17–0.01
Na_2O	0.14–0.01
K_2O	2.85–2.31

fire retardants. As a result of its use and application, silica can be found in many places, including food ingredients such as pickles or pottery shards. Synthetic silica produced from silicon dioxide is modified to produce fibres that are stronger, thinner, and cheaper. In addition to its use in petrochemical and biochemical refining, silicon is also important in the synthesis of polymers, fibres, ceramics, and electronic components.

To utilize rice husk ash effectively, several steps are required to convert it into silica. Principally, there are two methods: burning and chemical reactions. These steps include drying, heating the rice husks until they become charcoal, and mixing them with KOH and HCl solutions, followed by precipitation and dissolution with various acids. This process can yield high-purity silica. Initially, the dried rice husks are heated for 4 hours at 700°C until they

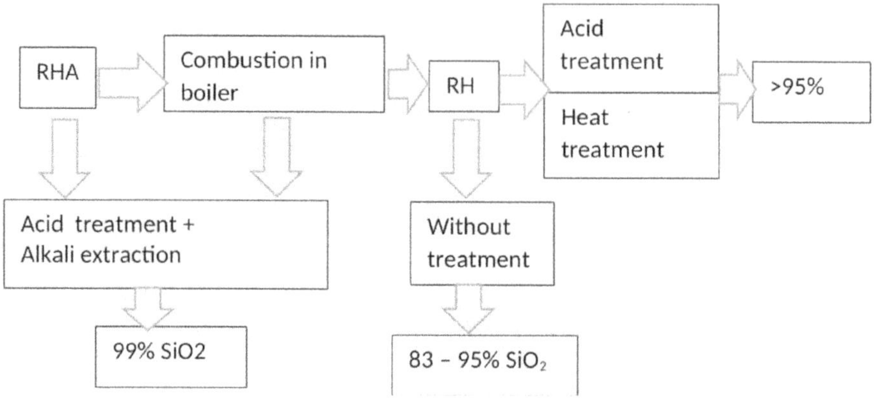

FIGURE 6.3
Processing ash husk rice [1].

turn into charcoal. Then, the ash is mixed with a 10% KOH solution at 85°C for 90 minutes [10]. After that, 1 M HCl solution was added slowly to the silicate solution until it reached pH 7 to form a precipitate. Next, the deposition process is carried out using various types of acids such as HCl, HF, HI, H_2SO_4, H_3PO_4, or $C_6H_8O_7$ [11]. This process will produce wet silica or silicon dioxide (SiO_2), where the oxidation process can convert wet silica into silicon [12, 13]. **Figure 6.3** illustrates the processing of rice husk ash into a silica mixture.

6.2 Rice Husk Applications

6.2.1 A Semiconductor

Rice husk ash is a significant source of alternative silica available in large quantities. To meet industrial standards, the quality of silica produced must undergo appropriate mechanical, physical, chemical, and thermal processes. One of the uses of silica from rice husk ash is as a semiconductor material for solar cells. Silica derived from rice husk ash offers several advantages over inorganic silica from quartz sand, including environmental friendliness, lower production costs, energy savings, and a simpler production process. Thus, it is very possible for silica from rice husk ash to be further developed and commercialized in the industry [13].

Silica from rice husk ash can be an alternative material for semiconductors used in solar cells, especially monocrystalline semiconductors. Semiconductors are materials that have a conductivity level between insulators and conductors, with a band gap energy (*Eg*) between 0 eV and 4 eV [14]. Solar cells are made from crystalline silicon from husk ash due to its abundant presence, non-toxic nature, and high conversion efficiency. Rice husk ash has the main component of amorphous silica with a content of around 83–90% [13]. To increase the efficiency of silica from husk ash, further development and additional treatments such as gold coating, further crystallization and oxidation are required. An example of the use of silica from rice husk ash has been applied to various types of solar cells, both monocrystalline solar cells and the next generation of solar cells that are not silica-based, such as *dye-sensitized solar cells (DSSC)* [15].

6.2.2 RHA as Composite Particles

Research using RHA as an additional particle in resin composites of up to 20% has increased the fracture toughness of epoxy by 157%. This toughness increases to 25% when using 10% rice husks. When adding RHA above 20%, the fracture toughness begins to decrease but is still higher than pure epoxy. An increase in composite fracture toughness of 3.12% occurred when using 20% by weight of rice husk. Mixing rice husk particles with epoxy significantly increases the adhesion of rice husk to the resin. In addition, hybridization with rice husk ash increased fracture toughness by 7.15% compared to samples without RHA added. This shows that rice husk ash is a suitable component for producing biocomposites. RHA composition has a significant influence on the flexural stress of the composite. **Figure 6.3** illustrates how variations in RHA composition affect the flexural stress in the composite.

The addition of RHA to epoxy significantly increases the glass transition temperature (Tg) of the composite and has better thermal stability than pure epoxy. From experiments [16], it can be seen that the coefficient of thermal expansion (CTE) of epoxy resin was reduced by 29.2% after the addition of 18.81 vol.% RHA. Deposit modulus composite epoxy/RHA increases significantly up to 55.3%. Compared to epoxy pure, composite with 18.81 vol.% RHA shows enhancement violence micro amounting to 44.7%.

The influence of RHA composition on the flexural strength of the resin composite is shown in **Figure 6.4**. The figure illustrates that the addition of RHA from 5% to 20% results in the highest flexural strength at a 10% addition. Further increases in RHA content beyond 10% lead to a decrease in flexural strength.

6.2.3 RHA as Mixture Concrete Composite Reinforced

Currently, RHA is widely used as a mixture in construction materials due to its porous structure and excellent insulating properties. [18]. Profit use of material silica in concrete reduces porosity, increases density, and improves the *interfacial transition zone* (*ITZ*) between cement and aggregate [19]. After being transformed into charcoal and mixed into concrete, rice husk ash

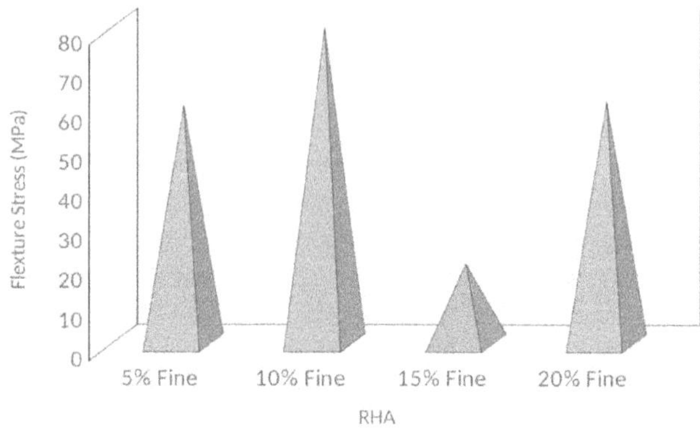

FIGURE 6.4
Influence RHA composition against voltage flexture resin composites [17].

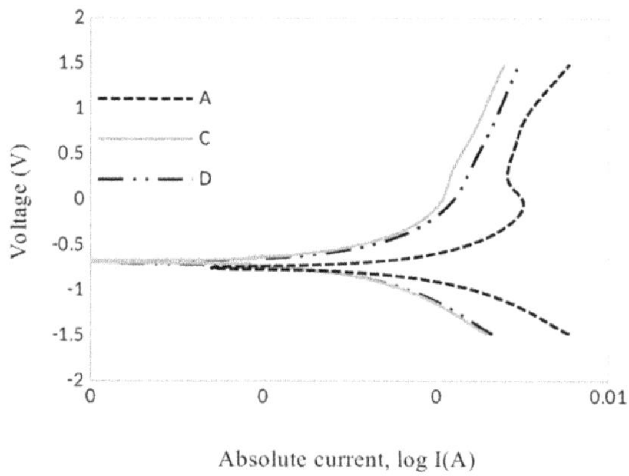

FIGURE 6.5
Effect of silica content on the polarization diagram of iron reinforcement in concrete after immersion in NaCl for 60 days. (A: no silica, C: 30%, D: 50%) [20].

(RHA) applications not only enhance the strength of structural concrete but also improve the pitting resistance of iron, as demonstrated in **Figure 6.5**. The use of silica in concrete significantly enhances the corrosion resistance of iron reinforcement. Concrete without silica (A) exhibits a higher level of corrosion compared to concrete that contains silica. Specifically, concrete with a silica content of 50% (C) demonstrates increased resistance to corrosion; however, concrete with a silica content of 30% (D) shows the most optimal resistance. This indicates that an appropriate increase in silica content in concrete can substantially reduce the rate of corrosion in iron reinforcement.

Composite concrete incorporating rice husk ash (RHA) filler has been shown to enhance the mechanical properties of concrete, including increases in the modulus of elasticity (MOE), compressive strength, and flexural strength. The MOE of the blended concrete rises with a higher proportion of RHA

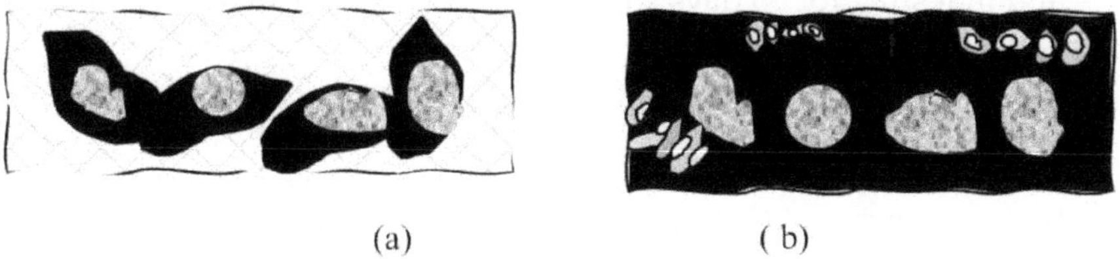

FIGURE 6.6
Pozzolanic reaction of RHA [22]. a: before reaction pozzolaic , b: after pozzolanic reaction.

substituted for ordinary Portland cement (OPC). This improvement is attributed to the fine particles of RHA, which effectively fill the pores within the matrix, leading to a denser gel matrix. Furthermore, the increase in strength is linked to the high silica (SiO_2) content in the ash, which reaches 91.13%. This silica reacts with calcium oxide (CaO) in the cement during the hydration process, facilitating the formation of C3S and C2S bonds that enhance the binding and integration of the existing aggregates. Additionally, the fine silica particles contribute significantly by filling voids within the concrete matrix, resulting in a denser and more durable material [21]. The pozzolanic reaction of rice husk ash (RHA) is illustrated in **Figure 6.6**. It demonstrates the interactions between RHA and calcium hydroxide resulting from the hydration of cement, which produces *calcium silicate hydrate* (CSH) that strengthens the concrete. This figure also shows the changes in the microstructure of concrete as a consequence of the pozzolanic reaction, leading to increased strength and durability of the concrete.

Apart from improving the characteristic mechanics of concrete, a number of benefits of using RHA concrete are as follows [22]:

- Reduce shrinkage drying: during the drying process, concrete experiences shrinkage as water evaporates from the concrete matrix along with the hardening process. It was recorded that concrete without rice husk ash (RHA) exhibited a deformation of approximately 600 μm/m. However, this deformation was reduced by about 50% when 10% RHA was added.
- Permeability: permeability refers to the property of concrete that allows water and aggressive chemical materials to penetrate its matrix, leading to various chemical attacks such as carbonation and chloride ion intrusion. These processes can ultimately result in the corrosion of iron reinforcement. The permeability of concrete is closely related to the durability and corrosivity of iron reinforcement. Test results indicate that when rice husk ash (RHA) is used in self-healing concrete, increasing the amount of RHA leads to a significant reduction in permeability. The RHA concentration was 55%,

and in testing for 56 days showed a mark decline in permeability of 1.10^{-10} m/s [23].

- Water absorption and sorptivity: the quality of concrete is influenced by its water absorption and sorptivity properties. Therefore, it is essential to control these factors to prevent the intrusion of harmful ions (such as $SO4^{2-}$ and Cl^{-}) from the environment, which can damage and compromise the structural integrity of concrete components. Generally, the porosity of concrete is reduced due to the pore refinement capabilities provided by rice husk ash (RHA) particles. The water absorption of the RHA mixture decreases primarily due to the increase in gel density, reduction in pore size, and decrease in void content. This improvement continues until the RHA content exceeds 20%.
- Chloride penetration: corrosion is the cause of damage to iron reinforcement in concrete, especially in reinforced concrete structures used in infrastructure, due to the entry of chloride elements in the concrete. It was observed that a reduction in permeability, chloride penetration, and chloride diffusion of about 35%, 75%, and 28% respectively was observed when 30% RHA was used. This study showed that the addition of 15% RHA particles could increase resistance to chloride ion penetration by approximately 4 times compared to the control mixture for seven days [22].
- Resistance to alkali-silica reaction: the alkali-silica reaction (ASR) poses a significant challenge to concrete quality. Experiments demonstrated that the incorporation of rice husk ash (RHA) effectively resists ASR for durations of 16 and 30 days. At 175 days, it was observed that mixtures containing 7% and 10% RHA exhibited reductions in ASR expansion of approximately 52% and 33%, respectively, compared to concrete made with ordinary Portland cement (OPC) alone [22].
- Resistance to carbonation: carbon dioxide (CO_2) from the atmosphere can penetrate the concrete matrix and react with its basic components, particularly calcium hydroxide ($Ca(OH)_2$), leading to potential damage to the concrete structure. This carbonation process reduces the pH within the pore system, and when the alkalinity of the concrete is compromised, the reinforcing steel becomes more susceptible to corrosion. Studies have shown that rice husk ash (RHA) is more effective than fly ash (FA) in enhancing corrosion resistance in concrete mixtures.
- Electrical and thermal conductivity: electrical resistivity is a quantity used to evaluate the quality of concrete structures. It was found that the electrical conductivity increased if the RHA content increased from 10% to 20%. The increase in electrical resistivity

reached 340–404%. The actual electrical resistivity of concrete with RHA content between 0 and 20% is between 5 and 10 k Ω cm. With an increase in the number of RHAs, the resistivity is recorded to increase by approximately 52% [22].

- Resistance to acid and sulphate attack: chemical elements such as SO_4 ions produced from sulphate sources ($MgSO_4/H_2SO_4$) can react with the presence of $Ca(OH)_2$ in concrete (produced from gypsum or during hydration), which facilitates volumetric expansion and the separation of concrete components/materials from their original mass. It was noted that the addition of RHA in concrete resulted in better acid attack resistance compared to OPC due to the presence of reactive silica in the RHA particles [23]. This activity of RHA concrete was observed as a result of reduced penetration and lower $Ca(OH)_2$ content in RHA-containing cement concrete, leading to increased transition zones and reduced porosity. However, the compressive strength of RHA concrete decreases by around 5–16% when the specimen is subjected to a 5% $MgSO_4$ solution.

6.3 Industrial Process

6.3.1 RHA as Cooler Nano-Based

In the form of nano-sized particles, silica possesses advantages that make it highly versatile and widely used in various fields, including ceramics, materials chemistry, catalysis, chromatography, energy, electronics, coatings, stabilizers, emulsifiers, and biological sciences [3]. Large-scale industries require substantial quantities of silica nanoparticles with precise shape, size, and porosity [24]. The production of nanoparticles can be achieved through several methods, including vapour-phase reactions, sol-gel processes, and thermo-decomposition [25]. Among these methods, chemical synthesis is the most efficient, as it allows for easy control over material size, shape, and purity. However, for industrial applications, cost-effectiveness and the availability of large quantities of starting precursors are essential. Nanoparticles produced from rice husks present a promising alternative to silica sourced from natural materials.

Making nano-sized RHA is done by extraction. Pure nano-silica is extracted from rice husk ash (RHA) by controlling various process parameters. The process begins by washing the rice husks to remove dirt, then drying and burning them at a temperature of 973K for 3 hours in an inert atmosphere. The resulting RHA is then washed to neutralize the pH and remove impurities. After that, the RHA is refluxed with HCl to remove metal impurities and extract pure nano-silica. The process continues by boiling the filtered

RHA in a NaOH solution at a temperature of 353K for 1.5 hours while magnetically stirring, and then filtering to obtain sodium silicate. This sodium silicate is then precipitated into nano-silica by adding concentrated $H_2 SO_4$ to lower the pH. The obtained nano-silica precipitates are washed, filtered, and sintered at temperatures of 973 and 1373K for 3 hours in a muffle furnace. Finally, the nano-silica is ground to achieve the desired grain size [26].

Research into the potential use of RHA as a nanoparticle-based refrigerant has provided an environmentally friendly and cost-effective refrigerant alternative that is suitable for use in various industrial applications. The nanoparticles formed are able to increase cooling efficiency, reduce heat generated by the engine, and extend component life. Experiments using silica from rice husk ash (RHA) as a nanoparticle-based coolant showed positive results [25]. Silica is synthesized from RHA chemically and mixed into water at various concentrations, such as 0.10 g/L, 0.20 g/L, 0.25 g/L, 0.50 g/L, 0.75 g/L, 1.00 g/L, 1.25 g/L, 1.50 g/L, 2.00 g/L, and 2.50 g/L. Through SEM tests, it can be seen that the silica nanoparticles formed have an almost spherical shape with an average diameter ranging from 16 nm to 46 nm. With the EDS technique, it is known that RHA contains the main elements Si and O with concentrations of 56.2% and 20%, respectively, as well as a clear main phase of SiO_2, and trace elements such as C, S, K, Na, Mg, Fe, and Ca [27].

After being used as a coolant in cutting machines, it was observed that the nanofluid derived from RHA performed relatively better compared to the coolant without nanoparticles. Nanofluid with a concentration of 2.00 g/L produced the best results for smoother finishes and faster machining times, providing cost savings in applications. This shows the great potential of RHA as a nanoparticle-based coolant in the machining industry, increasing machining efficiency and quality while reducing operational costs [27].

6.4 Rice Husk Ash Briquettes

Energy demand continues to increase due to human population growth and increased industrialization activities globally. Fossil fuels such as petroleum products, coal, and natural gas are the main energy sources that meet about 80% of global energy needs [28]. However, reliance on these non-renewable resources poses significant challenges, with their increased use leading to higher levels of pollution, global warming, and decreased public health. In 2018, global CO_2 emissions from fuel combustion reached 32.8 billion tonnes [29] and these concentrations will continue to increase unless substantial efforts are made to reduce emissions. Global warming is projected to increase to 3.2°C by 2100 [30], making urgent emissions reductions necessary. Therefore, efforts to decarbonize energy production and reduce the use of fossil fuels are essential. The efficient use of biomass and the application of appropriate technologies must be prioritized to overcome these challenges and ensure access to clean and sustainable energy.

Currently, agricultural waste is often burnt directly without optimizing energy efficiency or controlling air emissions, or it is left to decompose on agricultural land, potentially releasing greenhouse gases and polluting water sources [31]. Direct use of unprocessed biomass can cause problems during storage, transportation, handling, and processing. To overcome these challenges, various strategies have been developed to convert biomass into secondary fuels with better characteristics, including biomass densification. In line with efforts to save fossil fuels and switch to renewable energy sources, the importance of utilizing biomass as fuel has increased significantly. Biomass gasification, a process in which organic materials are burnt in closed reactors to produce energy, is crucial in this endeavour and has been implemented in various countries [32–34]. Currently, Cambodia uses around 55 biomass gasification power plants. Most of these are small and medium enterprises operating in the 200 to 600 kW power range, which replaces about 75% of diesel oil use [33]. Traditional biomass sources such as firewood, charcoal, and agricultural residues play an important role in meeting more than 90% of global rural energy needs [32]. Among these, rice husk (RH) and rice straw are emerging as very promising biomass sources due to their abundance and potential to produce bioenergy sustainably. These materials not only offer a renewable alternative to fossil fuels but also help reduce the environmental impact associated with energy production. Thus, utilizing biomass for energy purposes is an important step towards achieving global sustainability goals.

Charcoal from rice husks is formed through incomplete combustion when heated below its combustion temperature. This process produces high carbon charcoal, which is suitable for various uses such as cooking, heating, and electricity generation. Making carbon (carbonization) of rice husks makes it possible to produce charcoal briquettes in various forms. However, the weakness of rice husk charcoal is the relatively low heat output due to the low density and light weight in the carbonization furnace. To increase combustion efficiency, compaction is necessary [35]. Compaction of biomass can be done by making biomass into pellets or briquettes. The resulting briquettes can have a density up to ten times that of the original material [36]. This process increases the mass and energy of biomass per unit volume, reduces storage requirements, increases transportation efficiency, reduces particulate emissions per unit volume of material transported or burnt, and ensures uniform feeding to industrial equipment such as boilers, gasifiers, and domestic stoves.

Research shows that agricultural waste such as rice husks, corn cobs, and straw dregs can be effectively compressed into briquettes [37, 38]. According to research on the composition of briquettes with different biomass materials, materials with higher lignin, starch, or protein content are more easily compacted than those with higher cellulose content. This method has encouraged efforts to mix various types of biomasses to increase efficiency. For example, sawdust from Scots pine trees mixed with wheat straw produces more durable pellets, rice bran is used as a binder in making rice straw briquettes, and olive pulp mixed with fibrous paper mill waste produces more durable briquettes [39].

In general, the series of charcoal briquette production processes begins with the carbonization stage, where the raw materials are burnt in a rotary kiln at high temperatures for one week. After the combustion process is complete, the air outlet is closed and the gas is vented for 1–2 hours before closing the exhaust hole. After cooling for two weeks, the carbonized wood is crushed using a hammer or roller crusher into charcoal pieces less than 5 mm in size. The next step is drying, where the crushed charcoal is dried to a moisture content of around 15% to avoid problems during the briquette moulding process. The moulding process is carried out by compressing the raw material using a briquette machine by adding moisture, adhesive, a temperature of around 105°F (40°C), and pressure. After that, the briquettes are dried again at 275°F (135°C) for 3–4 hours to reduce the water content to around 5% before being packaged or stored for further use [40].

6.5 Rice Husk/Rice Husk Ash as a Source of Silica

Rapid development in industry and technology has increased the need for innovative new materials. Natural resources are limited and increasingly depleting, promoting the search for materials from alternative sources. One ongoing need is for ceramic materials, which possess superior properties such as chemical non-reactivity, thermal insulation, fire resistance, high-temperature endurance, and superconductivity. Due to these properties, ceramics have various applications in industry. In general, ceramics are utilized in the production of refractories, glass, tiles, sanitary ware, tableware, and electrical equipment [41].

The main component in producing ceramics is silica (SiO_2). As the demand for ceramic production continues to grow, the need for silica materials also increases. Most manufacturers use silica sand, gravel, sandstone, granite, quartz, and quartzite as sources of silica for ceramic production [42]. All of these sources of silica are found naturally in nature. Excessive use of natural raw materials can lead to environmental impacts and pollution. Therefore, it is essential to utilize waste to obtain silica. One alternative is the use of agricultural waste, specifically rice husk ash.

Waste materials from agriculture, such as rice husk waste or other plant residues, are considered to have great potential to meet the need for these new materials. RHA has been proven to be the most promising waste as a potential source of silica and can also address the problem of agricultural waste. In this context, research on the use of agricultural waste to produce ceramic materials promises to provide solutions that are both economically and environmentally sound [43].

In recent decades, researchers have increasingly explored the applications of silica derived from RHA in various fields, especially in the ceramics industry. The following are various industrial products that use silica from rice husk ash.

6.5.1 Refractories

Rice husk ash (RHA) has low thermal conductivity, making it suitable for use as a material in making refractories. Thermal conductivity in conventional refractory materials originates from atomic vibrations caused by phonons at low temperatures and photon conductivity at high temperatures in crystalline materials. Thanks to its amorphic silica content, RHA exhibits non-heat-conducting properties in refractories [44]. RHA-based refractories are prepared by mixing various flux components (as a binder), plasticizers (to impart plasticity to the RHA), and pore-forming agents (to increase the porosity of the final product) [45]. Air trapped within the refractory pores reduces conductivity and serves as a barrier to heat flow. Refractories are generally used as a secondary lining in furnaces or kilns in the foundry industry. In steel casting, RHA is used as a heat-resistant layer and as insulation for molten metal in ladles and moulds, preventing cracking of the refractory due to temperature differences of up to 200°C when casting metal. The low thermal conductivity of RHA causes gradual cooling of the steel and uniform solidification over its entire surface [46, 47].

6.5.2 Silicon Carbide

Ceramics, especially *silicon carbide*, have unique properties such as high hardness, heat resistance, and durability that metals do not have. This makes ceramics ideal for applications where precision and reliability are essential, such as in the machining and metal-forming industries. Its ability to maintain sharp cutting edges at high temperatures makes *silicon carbide* a very reliable material for making cutting tools and abrasive media (grinding).

The presence of rice husks (RH) encourages a more economical and simpler SiC manufacturing process. RH has received considerable attention for its role in producing SiC fibres and particles. The use of RH in SiC production has encouraged further studies to industrialize this process through various improvements. Microwave technology has recently been used for SiC fabrication [48, 49], and researchers have explored various catalyst activities to enhance SiC formation [50]. Metal catalysts such as Fe, Ni, Cr, Co, and Pd significantly accelerate the reaction rate between carbon and SiO_2 at temperatures of 1200°C to 1600°C, resulting in high production rates and favourable particle morphology. The RH synthesis process at low temperatures and with economical raw materials makes it very attractive. Most studies focus on a two-step process for fabricating SiC from RH: first, removing volatiles from RH through a controlled atmosphere cooking process (400–800°C); and second, heating carbon-rich RH at high temperatures (>1300°C) to react with silica and form SiC [51]. The chemical reaction for the formation of SiC from RH can be summarized as the following 6.1–6.4 reactions [52]:

$$SiO_{2(S)} + C_{(S)} \rightarrow SiO_{(g)} + CO_{(g)} \tag{6.1}$$

$$SiO_{(g)} + 3CO_{(g)} \rightarrow SiC_{(S)} + 2CO_{2(g)} \tag{6.2}$$

$$SiO_{(g)} + 2C_{(S)} \rightarrow SiC_{(S)} + CO_{(g)} \tag{6.3}$$

$$CO_{2(g)} + C_{(S)} \rightarrow 2CO_{(g)} \tag{6.4}$$

Recently, *Li et al.* [53] introduced a new method to synthesize SiC nanowires from RH silica without a catalyst or protective atmosphere. They mixed RHA and phenolic resin in a high-speed planetarium mill at a stoichiometric ratio suitable for SiC production.

Apart from the products mentioned above, there are many other industrial applications that utilize silica derived from rice husk ash, starting from oxide and non-oxide products such as advanced ceramics, glass, whiteware, oxide ceramics: mullite ($3Al_2O_3.2SiO_2$), *cordierite* is a *magnesium aluminosilicate* ($Mg_2Al_4Si_5O_{18}$), *lithium aluminosilicate*, *forsterite*, (Mg_2SiO_4), *wollastonite* ($CaSiO_3$); non-oxide ceramics: C/SiO_2 composite, *silica airgel*, and *silicon nitride* [1].

References

1. Al-Khalaf, M. N., & Yousif, H. A. (1984). Use of rice husk ash in concrete. *International Journal of Cement Composites and Lightweight Concrete, 6*(4), 241–248.
2. Garboczi, E. J. (1990). Permeability, diffusivity, and microstructural parameters: A critical review. *Cement and Concrete Research, 20*(4), 591–601.
3. Hossain, S. S., Mathur, L., & Roy, P. K. (2018). Rice husk/rice husk ash as an alternative source of silica in ceramics: A review. *Journal of Asian Ceramic Societies, 6*(4), 299–313. https://doi.org/10.1080/21870764.2018.1539210
4. Aprianti, E., Shafigh, P., Bahri, S., & Farahani, J. N. (2015). Supplementary cementitious materials origin from agricultural wastes-A review. *Construction and Building Materials, 74*, 176–187.
5. Della, V. P., Kuhn, I., & Hotza, D. (2002). Rice husk ash as an alternate source for active silica production. *Materials Letters, 57*, 818–821.

6. Agrela, F., Cabrera, M., Morales, M., Zamorano, M., & Alshaaer, M. (2021). Biomass fly ash and biomass bottom ash. In *New trends in eco-efficient and recycled concrete* (pp. 23-58). Woodhead Publishing. https://doi.org/10.1016/B978-0-08-102480-5.00002-6
7. Raheem, A. A., & Kareem, M. A. (2017). Chemical composition and physical characteristics of rice husk ash blended cement. *International Journal of Engineering Research in Africa, 32,* 25–35.
8. Li, Y., Zhao, C., Ren, Q., Duan, L., Chen, H., & Chen, X. (2009). Effect of rice husk ash addition on CO2 capture behavior of calcium-based sorbent during calcium looping cycle. *Fuel Processing Technology, 90,* 825–834.
9. Srivastava, V. C., Mall, I. D., & Mishra, I. M. (2006). Characterization of mesoporous rice husk ash (RHA) and adsorption kinetics of metal ions from aqueous solution onto RHA. *Journal of Hazardous Materials, 134*(1–3), 257–267.
10. Agung, M., Hanafie, M. R., & Mardina, P. (2013). Ekstraksi silica dari abu sekam padi (Extraction of silica from rice husk ash). *Konversi, 2*(1), 28–31. [In Indonesian]
11. Kurama, H., & Kurama, S. K. (2003). *The effect of chemical treatment on the production of active silica from rice husk.* International Mining Congress and Exhibition of Turkey-IMCET, pp. 431–435.
12. Kwan, W. H., & Wong, Y. S. (2020). Acid leached rice husk ash (ARHA) in concrete: A review. *Materials Science and Technology, 3,* 501–507.
13. Putranto, A. W., Abida, S. H., Sholeh, A. B., & Azfa, H. T. (2021). The potential of rice husk ash for silica synthesis as a semiconductor material for monocrystalline solar cell: A review. *IOP Conference Series: Earth and Environmental Science, 733.* https://doi.org/10.1088/1755-1315/733/1/012029
14. Kurama, H., & Kurama, S. K. (2003). The effect of chemical treatment on the production of active silica from rice husk. *International Mining Congress and Exhibition of Turkey-IMCET,* 431–435.
15. Pode, R. (2016). Potential applications of rice husk ash waste from rice husk biomass power plant. *Renewable and Sustainable Energy Reviews, 53,* 1468–1485.
16. Darekar, V. S., Kulthe, M. G., Goyal, A., et al. (2024). Rice husk ash: Effective reinforcement for epoxy-based composites for electronic applications. *Journal of Electronic Materials, 53,* 1344–1359. https://doi.org/10.1007/s11664-023-10835-7
17. Shehab. (2022). *Development of waste rice husk ash and polyester composite.* Final Year Project, Inti International University.
18. Choi, N. W., Mori, I., & Ohama, Y. (2006). Development of rice husks-plastics composites for building materials. *Waste Management, 26*(2), 189–194
19. Ye, G., Huang, H., & van Tuan, N. (2018). *Rice Husk Ash.* Springer: Berlin, Germany.
20. Asmara, Y. P., Siregar, J. P., Tezara, C., Nurlisa, W., & Jamiluddin, J. (2016). Research article: Long term corrosion experiment of steel rebar in fly ash-based geopolymer concrete in NaCl solution. *International Journal of Corrosion,* 1–6. https://doi.org/10.1155/2016/3853045
21. Hidayat, A. (2011). Pengaruh Penambahan Abu Sekam Padi Terhadap Kuat tekan beton K-225. *E-Journal Aptek, 3,* 162.
22. Amran, M., Fediuk, R., Murali, G., Vatin, N., Karelina, M., Ozbakkaloglu, T., Krishna, R. S., Sahoo, A. K., Das, S. K., & Mishra, J. (2021). Rice husk ash-based concrete composites: A critical review of their properties and applications. *Crystals,* 11(2), 168. https://doi.org/10.3390/cryst11020168

23. Dwi Beauty Ratnawuri Hanafi. (2014). *Self healing capability beton dengan persentase fly Ash 0%, 20%, 25%, 30%, 35%, 45% Dan 55% Sebagai Pengganti Sebagian Semen Ditinjau Dari Workability, Kuat Tekan Dan Permeabilitas*. Fakultas Keguruan Dan Ilmu Pendidikan Universitas Sebelas Maret Surakarta.
24. Yuvakkumar, R., Elango, V., Rajendran, V., & Kannan, N. (2012). High-purity nano silica powder from rice husk using a simple chemical method. *Journal of Experimental Nanoscience, 9*(3), 272–281. https://doi.org/10.1080/17458080.2012.656709
25. Wang, Z. L., Gao, R. P., Gole, J. L., & Stout, J. D. (2000). Silica nanotubes and nanofiber arrays. *Advanced Materials, 12*(19), 1938–1940.
26. Tomozawa, M., Kim, D. L., & Lou, V. (2001). Preparation of high purity, low water content fused silica glass. *Journal of Non-Crystalline Solids, 296*, 102–106.
27. Afolalu, S. A., Egbe, M., & Emetere, M. E. (2021). Development and performance evaluation of silica nano-cutting fluids from rice husk ash (RHA) for metalworking and machining operations. *Journal of Bio- and Tribo-Corrosion, 7*, 146. https://doi.org/10.1007/s40735-021-00582-9
28. Kpalo, S. Y., Zainuddin, M. F., Manaf, L. A., & Roslan, A. M. (2020). A review of technical and economic aspects of biomass briquetting. *Sustainability, 12*(11), 4609. https://doi.org/10.3390/su12114609
29. International Energy Agency. (2019). *CO2 emissions from fuel combustion—highlights*. Retrieved November 20, 2019, from www.iea.org/tandc/
30. United Nations Environment Programme. (2019). *Emissions gap report*. Retrieved November 20, 2019, from https://www.unenvironment.org/resources/emissions-gap-report-2019
31. Muazu, R. I., & Stegemann, J. A. (2015). Effects of operating variables on durability of fuel briquettes from rice husks and corn cobs. *Fuel Processing Technology, 133*, 137–145.
32. Pode, R. (2016). Potential applications of rice husk ash waste from rice husk biomass power plant. *Renewable and Sustainable Energy Reviews, 53*, 1468–1485.
33. Binod, P., Sindhu, R., Singhania, R. R., Vikram, S., Devi, L., Nagalakshmi, S., Kurien, N., Sukumaran, R. K., & Pandey, A. (2010). Bioethanol production from rice straw: An overview. *Bioresource Technology, 101*(13), 4767–4774.
34. Matsumura, Y., Minowa, T., & Yamamoto, H. (2005). Amount, availability, and potential use of rice straw (agricultural residue) biomass as an energy resource in Japan. *Biomass and Bioenergy, 29*(5), 347–354.
35. Kaliyan, N., & Morey, R. V. (2009). Factors affecting strength and durability of densified biomass products. *Biomass and Bioenergy, 33*(3), 337–359.
36. Vassilev, S. V., Baxter, D., Andersen, L. K., & Vassileva, C. G. (2010). An overview of the chemical composition of biomass. *Fuel, 89*(5), 913–933.
37. Vadiveloo, J., Nurfariza, B., & Fadel, J. G. (2009). Nutritional improvement of rice husks. *Animal Feed Science and Technology, 151*(3–4), 299–305.
38. Williams, P. T., & Nugranad, N. (2000). Comparison of products from the pyrolysis and catalytic pyrolysis of rice husks. *Energy, 25*(6), 493–513.
39. Tako, M., & Hizukuri, S. (2002). Gelatinization mechanism of potato starch. *Carbohydrate Polymers, 48*(4), 397–401.
40. https://www.ftmmachinery.com/blog/composition-and-processing-of-charcoal-briquette.html#:~:text=Step%201%3A%20Carbonization,exhaust%2C%20close%20the%20exhaust%20hole

41. Rahaman, M. N. (2003). *Ceramic processing and sintering* (2nd ed.). Marcel Dekker, Inc.
42. Lavender, M. D. (1999). The importance of silica to the modern world. *Indoor Built Environment, 8,* 89–93.
43. Zhu, M., Ji, R., Li, Z., et al. (2016). Preparation of glass ceramic foams for thermal insulation applications from coal fly ash and waste glass. *Construction and Building Materials, 112,* 398–405.
44. Pal, A. R., Bharati, S., Krishna, N. V. S., et al. (2012). The effect of sintering behaviour and phase transformations on strength and thermal conductivity of disposable tundish linings with varying compositions. *Ceramics International, 38,* 3383–3389.
45. Sobrosa, F. Z., Stochero, N. P., Marangon, E., et al. (2017). Development of refractory ceramics from residual silica derived from rice husk ash. *Ceramics International, 43,* 7142–7146.
46. Sembiring, S., Simanjuntak, W., Situmeang, R., et al. (2016). Preparation of refractory cordierite using amorphous rice husk silica for thermal insulation purposes. *Ceramics International, 42,* 8431–8437.
47. Sembiring, S., Simanjuntak, W., Situmeang, R., et al. (2017). Effect of alumina addition on the phase transformation and crystallization properties of refractory cordierite prepared from amorphous rice husk silica. *Journal of Asian Ceramic Societies, 5,* 186–192.
48. Li, J., Shirai, T., & Fuji, M. (2013). Rapid carbothermal synthesis of nanostructured silicon carbide particles and whiskers from rice husk by microwave heating method. *Advanced Powder Technology, 24,* 838–843.
49. Moshtaghioun, B., Poyato, R., Cumbrera, F., et al. (2012). Rapid carbothermic synthesis of silicon carbide nano powders by using microwave heating. *Journal of the European Ceramic Society, 32,* 1787–1794.
50. Krishnarao, R. (1995). Effect of cobalt catalyst on the formation of SiC from rice husk silica–carbon black mixture. *Journal of Materials Science, 30,* 3645–3651.
51. Zawrah, M., Zayed, M., & Ali, M. R. (2012). Synthesis and characterization of SiC and SiC/Si3N4 composite nano powders from waste material. *Journal of Hazardous Materials, 227–228,* 250–256.
52. Narciso-Romero, F., & Rodriguez-Reinoso, F. (1996). Synthesis of SiC from rice husks catalyzed by iron, cobalt or nickel. *Journal of Materials Science, 31,* 779–784.
53. Li, W., Huang, Q., Guo, H., et al. (2018). Green synthesis and photoluminescence property of β-SiC nanowires from rice husk silica and phenolic resin. *Ceramics International, 44,* 4500–4503.

7

Palm Oil: Renewable Material and Environmental Sustainability

7.1 Introduction

Oil palm (*Elaeis guineensis*) originates in West Africa, especially from the Guinea region [1]. This plant was introduced to Southeast Asia in the early 20th century, and has since become one of the world's most important agricultural commodities. Indonesia and Malaysia are the two main palm oil-producing countries, with extensive plantations spread throughout the two countries. Palm oil thrives in tropical climates with high rainfall and average temperatures between 24°C and 28°C. Oil palm plantations are generally located in the lowlands with an altitude of under 500 m above sea level. In Indonesia, areas such as Sumatra, Kalimantan, and Sulawesi are the main centres for oil palm cultivation and produce 49% of world production [1]. In Malaysia, palm oil plantations are distributed across the Malaysian Peninsula, Sabah, and Sarawak, accounting for 37% of global palm oil production. The country produces approximately 8.5 million tonnes of fresh fruit bunches annually. This production generates significant waste, including 5.4 million tonnes of fibre, 2.3 million tonnes of shells, and 8.8 million tonnes of empty fruit bunches each year [2]. The oil palm tree is composed of several key parts, including *fronds* (leaves), the *trunk* (stem), and *fruit bunches*. Each fruit within the bunches has a distinct structure: the *exocarp*, which forms the outer skin; the *mesocarp*, a fleshy layer abundant in palm oil; the *endocarp*, a hard shell that encases the seed; and the *kernel*, the seed core that serves as a valuable source of palm kernel oil. The components of the coconut palm are illustrated in **Figure 7.1**.

The coconut palm produces two main types of oil: palm oil, which is extracted from the fruit's mesocarp (meat), and palm kernel oil, which is derived from the seeds (*kernels*). Raw coconut oil is reddish in colour and contains approximately 50% saturated fatty acids (primarily palmitic acid), 40% monounsaturated fatty acids, and 10% polyunsaturated fatty acids. Additionally, it contains carotenoids (such as *β-carotene*), vitamin E, *and β-sitosterol* [3, 4].

DOI: 10.1201/9781003617594-7

FIGURE 7.1
Elements of coconut palm oil tree. Description: OPEFB: oil palm empty fruit bunches, OPMF: oil palm mesocarp fibre, OPKS: oil palm kernel shell (Illustration: canva, 2024, https://www.canva.com).

Palm oil is used in various products such as cooking oil, margarine, soap, cosmetics, and biodiesel fuel. Palm kernel oil is often used in the food and cosmetics industry because of its different fatty acid content. In addition, production waste such as empty fruit bunches, fibre, and shells are used as biomass fuel and organic fertilizer. Oil palm fruits are harvested from the bunches after they reach a certain maturity. Fresh fruit bunches (FFB) are steamed in a sterilizer to stop enzyme activity that can damage the oil. The fruit is separated from the bunch through a threshing process. The separated fruit is then processed in a pressing machine to extract oil from the fruit flesh. The oil obtained is then processed further to remove impurities and obtain pure palm oil. Palm kernel seeds (kernels) are dried and crushed to extract palm kernel oil.

The palm oil processing involves several key steps: fresh fruit bunches are cut from oil palm trees using specialized harvesting tools, transported to processing plants for further processing, and sterilized to facilitate the release of fruit while reducing water content. The fruit is then separated from the bunches using a threshing machine, followed by squeezing the separated fruit to extract oil from the flesh. The extracted oil is refined to remove impurities, resulting in clear and clean oil, which is then packaged for distribution to domestic and international markets [5]. Crude palm oil (CPO) can be produced using two extraction methods: wet and dry processes. In the wet method, the palm fruit is treated with steam and absorbs additional water before extraction, whereas in the dry method, the fruit is heated dry before undergoing the extraction process. These two methods produce different grades of oil. In general, the oil extraction procedure is carried out using organic solvents. Recently, alternative methods such as the formation of microemulsions in aqueous solutions containing surfactants and extraction with supercritical fluid (SFE) have also been used for oil extraction, with the aim of reducing the environmental impact of the use of organic solvents [1].

Palm oil has several components that are important for various applications and uses [6]. *Fresh fruit bunches (FFB)* contain a lot of palm fruit, which is harvested for oil extraction. *The oil palm fruit* consists of the outer skin (exocarp), flesh (mesocarp) which contains oil, and seeds (endocarp or kernel). *Palm oil kernels* contain palm kernel oil which is different from palm oil extracted from the pulp [2]. *Empty fruit bunches (EFB)* is a bunch that has had its fruit harvested and are often used as biomass fuel or returned to the garden as organic fertilizer. *Palm oil* leaves can be used as animal feed or compost material, while oil palm stems are usually not used in the oil industry but can be used as building materials or biomass fuel. *Oil palm* roots function to absorb nutrients and water from the soil. Additionally, *POFA (palm oil fuel ash)*, produced from burning biomass residues, has gained attention as a sustainable cement substitute in construction. The *OPKS (oil palm kernel shell)*, a hard by-product of palm kernel processing, serves as an eco-friendly alternative to coal in industrial boilers. Finally, *POME (palm oil mill effluent)*, a liquid waste from palm oil mills, is treated to mitigate its environmental impact and harnessed for biogas production.

Table 7.1 shows an analysis of the composition of palm oil solid residue. **Table 7.2** displays the composition of fresh palm fruit waste products (FFB). Palm oil empty fruit bunches (EFB) waste consists of fibre, skin, and fruit kernels remaining after the palm oil extraction process.

TABLE 7.1

Results of Analysis of Palm Oil Composition [2]

	Fibre	Shell	EFB
Volatile matter (wt %)	72.8	76.3	75.7
Fixed carbon (wt %)	18.8	20.5	17
Ash (wt %)	8.4	3.2	7.3

TABLE 7.2

Distribution of Products/Wastes from Fresh Fruit Bunch (FFB) [7]

Products/Waste	Percentage by Weight to FFB (Dry Basis)
Palm oil	21
Palm kernels	7
Fibres	15
Shell	6
EFB	23
Palm oil mill effluent (POME)	28
Total	100

7.2 Oil Palm Fibre-Reinforced Polymer Composites

With the increasing global effort to find more sustainable energy solutions, the utilization of materials with significant potential in the construction industry is gaining attention. The coconut palm, primarily cultivated for oil production, generates a substantial amount of biomass, particularly from empty fruit bunches (EFB), which are often underutilized or regarded as waste. The diverse applications and environmental benefits of natural fibres make them ideal materials for reinforcing composites. Among various types of natural fibres, palm fibre has received significant attention due to its abundant availability and good mechanical properties. *Lignocellulosic* fibre can be obtained from the stem, midrib, fruit mesocarp, and empty fruit bunches (TBO) of oil palm trees. TBO is the fibre mass remaining after the fruit is separated from the fresh fruit bunch (FFB). TBO can produce as much as 73% fibre [8]. Compared to wood, palm oil fibre has much thicker cell walls, which results in a much higher stiffness index, making it more stable [9] when used as reinforcing fibre. One example of empty fruit bunch fibre (EFB), namely *Elaeis guineensis,* has an average cellulose content of around 30–50%, hemicellulose and lignin of 15–35% and 20–30%, respectively [10]. This content is useful for narrowing the pores of concrete so that it is more resistant to cracking and chloride attack.

Polymer composites reinforced with palm fibre offer a promising solution for the development of sustainable materials. This composite utilizes the mechanical and thermal properties of palm fibre to improve the performance of the polymer matrix. The use of palm fibre in polymers not only reduces the environmental impact but also contributes to the efficient utilization of agricultural residues with better deformation properties than other natural fillers [11]. Producing composites with palm fibre is also less expensive compared to those made from thermoplastics. The following presents research on the use of palm fibre combined with resin, demonstrating its potential in creating high-performance and sustainable composites [12, 13].

The factors that can influence the mechanical properties of composites are tensile strength of fibres, adhesion between fibres and matrix, number of fibres/matrices, orientation of fibres, and their distribution in the matrix [14]. To increase the mechanical strength and facilitate their use in composites, chemical surface treatments such as the application of alkaline solutions, acids, and coupling agents are necessary. Based on the research results in **Table 7.3**, chemical treatment of palm oil fibres (OPFs) has been shown to improve compatibility, hydrophobicity, and interfacial bonding, thereby significantly improving the mechanical properties of these fibres.

The study conducted by Hussain [19] investigated the use of polypropylene-based oil palm empty fruit bunches (POEFB) as a reinforcing material with a 6061-O aluminium matrix. The composition of POEFB was varied by 10%, 20%, 30%, and 40%. The results showed that the tensile strength increased with increasing fibre content, reaching the highest tensile strength of 65.99 MPa at 40% fibre content. This study concluded that composite failure was caused by fibre/matrix interfacial bonding problems and filler-filler interactions at non-uniform fibre content. Additionally, delamination occurs due to improper surface preparation and contamination, resulting in poor bonding [20].

TABLE 7.3

Treatment Chemistry for Repair Characteristic OPFChemical [7]

Method	Conditions	Mechanical Properties Improvement	Ref.
Alkaline	1% and 5% NaOH for 2 hours at room temperature	5% increase in wheat gluten with 5% NaOH	[15]
Silane	2% triethoxy (ethyl) silane for 3 hours	120% compared with without treatment.	[16, 17]
Peroxide	4% H_2O_2 for 3 hours (ratio fibre to H_2O_2 = 1:20)	Enhancement highest in OPF with 4% H_2O_2 reached 6.32 MPa.	[16, 17]
Isocyanate	20% Poly [methylene (polyphenyl isocyanate)] (PMPPIC)	Increases flexural strength 20% compared to not treated.	[17, 18]

Research on the effect of palm fibre, *oil palm fibre* (OPF), with variations in treatment and matrix on mechanical strength (tensile strength, fracture energy, flexural strength, tensile modulus) has been carried out by various researchers as follows. The palm ash composite, *oil palm ash* (OPA), with an epoxy matrix showed impressive results with a flexural strength reaching 120.0 MPa, a tensile modulus of 5.9 GPa, a tensile strength of 61.6 MPa, and an impact modulus of 3.3 kJ/m^2 [21]. Furthermore, *oil palm empty fruit bunches* (EFB), which were combined with woven kenaf using the hand layup technique with epoxy as a polymer matrix, showed a flexural strength of 115.8 MPa, a tensile modulus of 8.7 GPa, a tensile strength of 55.7 MPa, and a fracture energy of 1.78 J [22]. Composites made from palm oil empty fruit bunches (EFB) combined with Kevlar in an epoxy matrix, pressed for 24 hours and exposed to gamma radiation, exhibited a flexural strength ranging from 41.2 to 66.9 MPa. The tensile strength was measured at 33.0 MPa, with a tensile modulus ranging from 4.4 to 5.7 GPa. Additionally, the impact strength was recorded at 1.6 kJ/m^2 [23].

Recent studies have thoroughly investigated various composites made from empty fruit bunches (EFB) to evaluate their mechanical properties under different manufacturing and treatment conditions. For instance, a composite consisting of 5% by volume of EFB fibres (10–20 mm long) used to reinforce an epoxy matrix exhibited a flexural strength of 40.9 MPa, a tensile modulus of 3.2 GPa, and a tensile strength of 29.9 MPa [24]. In contrast, composites incorporating 50% by weight of palm ash (0.28 mm particle size) in a urea-formaldehyde (UF) matrix demonstrated significantly lower mechanical performance, with a flexural strength of 1.4 MPa, a tensile strength of 3.9 MPa, and an impact modulus of 1.2 kJ/m^2 [25]. Additionally, the incorporation of 40% by weight of EFB fibres (40 mm in length) into a phenol-formaldehyde (PF) matrix, which underwent various chemical treatments such as mercerization and acrylation, resulted in a range of flexural strengths from 16 to 75 MPa and tensile moduli between 0.7 and 3.9 GPa. This indicates the flexibility of EFB composites under different treatment conditions [26].

7.3 Thermal Insulation

The construction industry has experienced rapid growth in recent years [27], leading to increased energy consumption and contributing to approximately 40% of total greenhouse gas emissions [28]. Experts predict that by 2030, carbon emissions from this sector will reach 40 billion tons, consuming 65% and 42% of total energy in the United States and the European Union, respectively. It is estimated that by 2035, around 75% of total energy will still come from fossil fuels [29]. Implementing appropriate energy efficiency practices can significantly reduce energy consumption, making it essential to explore strategies for efficient energy use to preserve nature.

Furthermore, raising awareness and implementing strategies for green building concepts aimed at reducing GHG emissions and conserving fossil fuels is crucial. Various initiatives have been proposed to identify and utilize sustainable building materials. Much of the energy consumption in building construction is related to the efficiency of air conditioning (HVAC) systems and thermal insulation materials, both of which influence overall thermal performance. Choosing construction materials with minimal environmental impact is essential for sustainable development. Currently, researchers are focusing on finding biodegradable, recyclable, low-cost, safe, and natural materials. Efficient thermal insulation is vital for reducing electricity demand by minimizing thermal conductivity [30]. The use of palm oil waste, particularly the fibre component, shows significant potential as a thermal insulation material. By utilizing palm fibre, we can not only minimize waste but also develop efficient insulating materials.

Insulating materials have been developed from organic, inorganic, metal, and waste sources, focusing on their thermal properties. Current research is directed towards using natural fibres with high lignocellulose content, making them suitable for wall insulation. Studies indicate that agricultural lignocellulosic byproducts and natural fibres can be used to produce bio-based insulating concrete. Examples include hemp, flax, sunflower, and date palm fibres, which have been researched for their insulating properties. Research has shown that the density of insulation boards affects thermal conductivity, with lower-density materials exhibiting lower thermal conductivity [31–33].

Globally, biomass products such as palm oil, wood, rice husks, bagasse, and coconut fibre generate significant waste [31]. These organic and natural byproducts have the potential to be used as insulation products in various industrial sectors. This review discusses the thermal, physical, and mechanical properties of oil palm empty fruit bunch fibre (OPEFB) and sugarcane bagasse fibre, focusing on their potential as efficient thermal insulators. The properties of OPEFB and bagasse fibres, as well as their composites, are discussed in detail.

Research has documented the thermal conductivity values of various natural fibre insulation materials, demonstrating good thermal performance. The thermal conductivity properties of agricultural wastes, such as empty oil palm fruit bunches, coconut fibre, and sugarcane bagasse mixed with clay, have been investigated [34]. Materials with higher conductivity allow heat to flow more quickly. Results show that test objects containing 2.5%, 5%, 7.5%, and 10% OPEFB fibre exhibited low thermal conductivity at 10% OPEFB, measuring 0.27 W/(m·K). This indicates that a higher fibre composition correlates with lower thermal conductivity.

The addition of materials such as clay has proven effective in enhancing insulating properties. In one study, the use of 5% OPEFB filler resulted in excellent improvements in thermal insulation properties [35]. Another researcher, Erwisyah [36], created insulation boards using OPEFB fibres, yielding thermal conductivity values between 0.049 and 0.054 W/(m·K) at a density of 116.4 kg/m³. Thermal conductivity values ranging from 0.2 to 0.069 W/(m·K) were achieved with materials at densities of 66 to 110 kg/m³ and temperatures from 40°C to 70°C, indicating that palm oil waste outperforms rock wool and glass fibre at low temperatures [37].

Agrawal [38] used phenol-formaldehyde composites reinforced with OPEFB fibres in various proportions (20–50% by weight), and the results demonstrated that the addition of OPEFB fibre significantly reduced thermal conductivity and diffusivity to 0.24 W/(m·K), much lower than pure phenol-formaldehyde, which has a thermal conductivity of 0.348 W/(m·K).

7.4 Geopolymer Concrete Using Palm Oil Fibre

Concrete is an indispensable construction material due to its high compressive strength and stiffness. However, its brittle nature under tensile stress necessitates reinforcement. The tensile strength of concrete can be enhanced by incorporating short fibres that are randomly distributed within the concrete mix. This addition helps prevent crack propagation and improves flexibility. In efforts to preserve nature, the use of natural fibres in concrete has been extensively tested. Natural fibres not only reduce the effects of drying shrinkage but are also cost-effective, environmentally friendly, and readily available. Potential fibres for reinforcing concrete include jute, hemp, sisal, ramie, coconut, and oil palm fibres. Currently, fibres from oil palm trunks (OPTF) are being utilized to address the millions of tons of biological waste generated each year that can be repurposed in the concrete industry.

Research indicates that using oil palm clinker aggregate (POCA) and coconut palm shells can produce lightweight structural concrete with strengths exceeding 25 MPa, while ash from burnt oil palm can yield high-strength concrete exceeding 80 MPa [39]. The stems, midribs, and empty fruit bunches of coconut palm provide fibres with significant tensile strength (300–600 MPa) and a density ranging from 500 to 1200 kg/m³, making them valuable for enhancing the mechanical characteristics of concrete [40].

A study on the influence of coconut palm fibre on the slump value was conducted to investigate the concrete's flowability when poured. The slump value serves as an indicator of the workability of concrete, reflecting its ability to flow without additional aid. A higher slump value signifies greater flexibility. In this research, the focus was on using palm oil clinker (POCA) and oil palm trunk fibre (OPTF). The slump value exhibited a linear decrease with increasing POCA content; a 25% increase in POCA content resulted in a 21% decrease in slump [38]. To enhance the slump value, a geopolymer binder was utilized with the addition of a naphthalene-based plasticizer (SP).

The inclusion of SP aimed to increase the viscosity of the concrete mixture [41].

Regarding slump values, the incorporation of SP at all percentages produced a stiffer mixture and improved cohesiveness. The effect of SP was also reflected in the compaction factor of the reference specimen, which increased by 7.14% for an SP content of 0.3% and by 9.52% for an SP content of 0.5%. The effect of SP varied based on fibre content, with compaction factor reductions of 8%, 17%, and 23% for fibre contents of 1%, 2%, and 3%, respectively [38].

Increasing the POCA content resulted in a significant decrease in strength, particularly in mixtures with 75% and 100% POCA, where strength diminished notably within the first 24 hours due to reduced workability at higher replacement ratios [39]. This decline in strength is attributed to the combined effect of increased POCA content and decreased workability, leading to a stiffer mixture with less effective compaction and more voids in the binder, as well as in the interface zone between the aggregate and the geopolymer binder. Furthermore, the higher viscosity of the geopolymer binder leads to more air voids being trapped within the binder matrix. This is visibly evident on the crack surface of the POCA-100 sample, where air voids ranged from 0.5 to 6 mm in diameter. The formation of these voids is related to the geopolymerization process, which occurs more rapidly than the hydration of ordinary cement [42].

Currently, research utilizing palm oil fly ash (POFA) and fly ash (FA) from agro-industrial waste has been conducted [43]. In the geopolymer concrete experiments, both POFA and FA were activated with a combination of sodium hydroxide and sodium silicate and then steam-dried for 24 hours at 60°C. The results indicated that POFA and local FA, as geopolymer source materials, could produce a mixture with a compressive strength ranging from 19 to 22.5 MPa on the 28th day.

The compressive strength, tensile strength, flexural strength, and elastic modulus of both types of geopolymer concrete tended to increase over time. The density of the FA and POFA geopolymer concrete was within the range typical of normal concrete [44]. Interestingly, the tensile strength of the concrete made with POFA was found to be higher than that of the FA geopolymer concrete. Conversely, FA geopolymer concrete exhibited higher flexural strength compared to POFA. This behaviour can be attributed to the good bonding between the aggregate and the geopolymer paste, which is enhanced by the high silicate content in the FA mixture [45].

Moreover, the modulus of elasticity of FA geopolymer concrete was higher than that of POFA. The high porosity and water absorption properties of POFA ash significantly reduce the elastic modulus compared to FA geopolymers [46]. Overall, the geopolymer concrete produced from local FA and POFA demonstrated good mechanical properties.

7.5 Applications of Palm Oil

7.5.1 Brick Mix

With high population growth, building and infrastructure construction will also increase, so it is necessary to increase brick production. Tata stone is considered an effective material for building infrastructure because it is cheap and fast in construction. However, traditional brick-making methods consume a lot of clay, with an estimated 3.13 billion m^3 of clay required per year [47]. The use of clay as a material for making bricks can cause environmental degradation and damage to nature. To reduce demand for clay, researchers are exploring alternatives such as using waste as a material for making bricks. Currently, trials are being conducted on the use of fly ash, mining waste, rice husk ash (RHA), and bagasse ash (SBA) as brick fillers [48].

Palm oil fuel ash (POFA) is a residual material produced from burning solid palm oil waste (such as palm kernel shells, mesocarp fibre, and empty fruit bunches) as boiler fuel to produce electricity in palm oil mills or power plants. As much as 5% of burnt solid waste is converted into POFA [49]. POFA is usually discarded in open fields due to its low nutritional value. Therefore, using POFA waste as a raw material in brick production may offer a solution to the POFA disposal problem, while reducing clay consumption in brick-making.

The addition of palm oil fly ash (POFA) has been shown to reduce brick shrinkage. Shrinkage occurs as a consequence of water evaporation from the bricks during the firing process. Experiments indicate that incorporating POFA decreases shrinkage; without POFA, bricks shrink by 4.19%, whereas those containing POFA experience shrinkage between 2.94% and 3.73% [50].

Another effect of using POFA is the reduction in weight per unit area of the bricks. Bricks without POFA have a weight of 76.41 kg/m^2, while the weight per unit area decreases by 10.76%, 12.53%, 18.68%, and 23.34% for bricks containing 5%, 10%, 15%, and 20% POFA, respectively [50]. However, the addition of POFA results in a linear decline in compressive strength. Bricks without POFA exhibit a compressive strength of 6.44 N/mm^2, while those containing 5%, 10%, 15%, and 20% POFA show declines in compressive strength of 48.34%, 58.13%, 74.56%, and 80.47%, respectively. This decrease in compressive strength is attributed to changes in porosity, density, and pore size [51].

7.5.2 Palm Oil-Based Adhesive

The continuous growth of the construction and furniture industries has led to an increasing demand for adhesives. To promote environmental sustainability, there is a need to replace synthetic adhesives, which are often not eco-friendly, with more natural and sustainable options. One abundant and highly promising natural adhesive source is coconut palm oil. The coconut palm is a primary producer of crude palm oil (CPO) as its main product. However, the CPO production process also generates solid and liquid waste. One example of this solid waste is palm kernel cake (PKC), which is typically used as livestock feed. Recent technological innovations have shown that PKC also has significant potential as a standard bioadhesive material due to the proteins it contains.

The first step in extracting protein from PKC involves grinding the cake and filtering it until it reaches a uniform size of 100 mesh [52]. The ground PKC is then suspended. For protein extraction, the PKC powder is mixed with 250 mL of a 55 mM NaOH solution at a ratio of 1:10 (g/mL). The PKC and NaOH suspensions are stirred for 24 hours at 25°C and 60°C. After this period, the suspension is centrifuged at 10,000 rpm and 4°C for 15 minutes. Hydrochloric acid (HCl) is then used as a reagent for protein precipitation. The protein content obtained at an extraction temperature of 25°C is 20.731%, while at 60°C, it is 28.152% [54].

Using particle board with a density of 0.798, the resulting elastic modulus, rupture modulus, and bond strength are 12 kg/cm², 125 kg/cm², and 4.5 kg/cm², respectively [49]. The amount of NaOH used also affects the shear strength of the particle board. Tests using 20% NaOH mixed with various biomass ratios (NaOH volumes: 1:3, 1:4, 1:5, 1:6, and 1:7) show that a 20% NaOH treatment at a 1:7 ratio yields the highest shear strength, measuring 0.32 ± 0.020 MPa [54].

In comparison to commercial adhesives, using adhesive derived from coconut palm oil on fibre boards demonstrates superior results, even with a lower adhesive content (**Table 7.4**). Additionally, the use of PKC adhesive represents a technological advancement in the fibre board industry with the potential to reduce emissions typically associated with urea-formaldehyde (UF) adhesives [55].

TABLE 7.4

Comparison of the Mechanical Properties of Commercial Adhesives and PKC Adhesives [55]

Properties	Commercial Adhesive	PKC-based Adhesive
Adhesive type	UV	PU
Tensile strength (Nm/m^2 .min)	1	2.4
Total adhesive (%)	12	12
E- Modulus (Nm/m^2 . min)	3,200	4,300
Bending strength (Nm/m^2 .min)	42	44
Swelling (% max)	10	7
Moisture content (% max)	9	6

7.5.3 Biofuels

The explosion in population, housing growth, and industrialization have significantly increased energy consumption and final production activities, leading to higher greenhouse gas (GHG) emissions, acid rain, and global warming. Therefore, efforts to reduce energy consumption and promote renewable energy are crucial for safeguarding the continuity of life on Earth. The development of biofuels from coconut palm is one step towards mitigating environmental impacts, especially considering that nearly 70% of global energy sources used in daily activities come from fossil fuels [56].

Biofuels can be categorized into three main types based on their physical state: solid, liquid, and gas. Solid biofuels include firewood, wood chips, wood pellets, and charcoal, all derived from biomass and widely used for heating and cooking needs. Liquid biofuels, such as bioethanol, biodiesel, bio-oil from pyrolysis, and drop-in biofuels, are produced through chemical processes and serve as alternative fuels for internal combustion engines. Meanwhile, gaseous biofuels consist of biogas and syngas, produced through anaerobic fermentation or biomass gasification, and are used to generate electricity or heat.

The use of biofuels in various forms offers a more environmentally friendly and sustainable energy solution compared to traditional fossil fuels [57]. Various biofuel products can be classified based on their physical state, as illustrated in **Figure 7.2**. This classification includes bioethanol, biodiesel, biogas, and biochar, each with distinct characteristics and applications in the renewable energy sector.

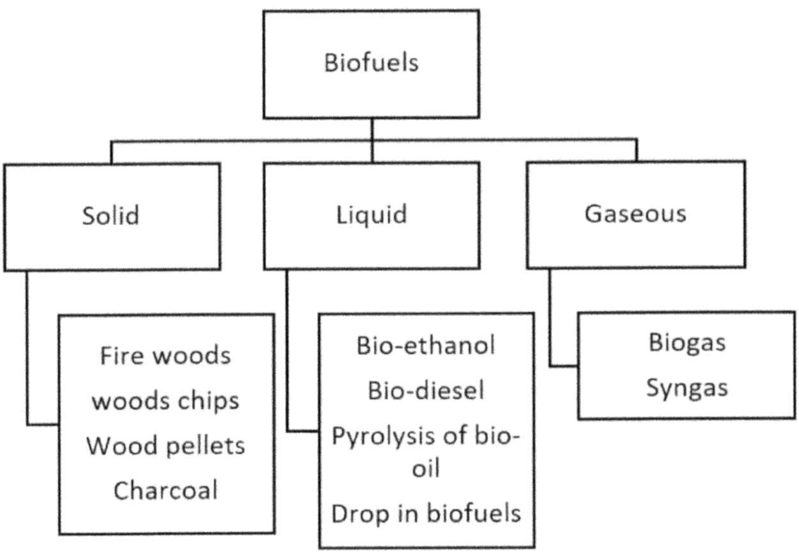

FIGURE 7.2
Classification of biofuels based on the shape.

Efforts to utilize coconut palm oil as a potential fuel source offer several advantages. Notably, its oil yield is significantly higher than that of other vegetable oil sources, reaching approximately 3.93 tons per hectare per year—almost triple the yield of its closest competitors. Additionally, coconut palm oil is more economical compared to other oils used in biodiesel production.

When compared to petroleum-based diesel, biodiesel derived from palm oil exhibits superior physical and chemical characteristics. For instance, it has a lower sulphur content and a reduced pour point, which facilitates handling in colder temperatures, making it suitable for use in cooler climates. Moreover, palm biodiesel generates less carbon residue, thereby reducing carbon buildup in diesel engines compared to petroleum diesel.

However, palm oil biodiesel also presents some physical and chemical drawbacks, including higher viscosity, a higher flash point, and lower gross heat of combustion [58]. The quality of biofuel can be enhanced by blending it with methyl esters or diesel oil. A series of experiments were conducted using samples of palm oil and palm oil methyl ester mixed with diesel at low stirring speeds. The palm oil and palm oil methyl ester were added in volume percentages of 5%, 10%, 15%, 20%, and 30%. The main properties, such as density, kinematic viscosity, and flash point, approached those of diesel fuel at a 30% fuel mixture [59].

References

1. Mohd Esa, N., Hosseini, S., Gangadaran, S., Lee, S. T., Kapourchali, F. R., & Moghadasian, M. H. (2013). Palm oil: Features and applications. *Lipid Technology, 25*(2), 25–48.
2. Husain, Z., Zainac, Z., & Abdullah, Z. (2002). Briquetting of palm fibre and shell from the processing of palm nuts to palm oil. *Biomass Bioenergy, 22*, 500–509. https://doi.org/10.1016/S0961-9534(02)00022-3
3. van der Vossen, H. A. M., & Mkamilo, G. S. (2007). *Plant resources of tropical Africa: Vegetable oils*. PROTA Foundation, Wageningen, Netherlands/Backhuys Publisher.
4. Wattanapenpaiboon, N., & Wahlqvist, M. L. (2003). Phytonutrient deficiency: The place of palm fruit. *Asia Pacific Journal of Clinical Nutrition, 12*(3), 363–368.
5. Kalam, M. A., & Masjuki, H. H. (2007). Experimental test of a diesel engine Envo-Diesel as an alternative fuel. In M. Chiaberge (Ed.), *New trends and development in automotive system engineering* (p. 649). Intech Malaysia.
6. Kawamura, F., Saary, N. S., et al. (2014). Subcritical water extraction of low-molecular-weight phenolic compounds from oil palm biomass. *Japan Agricultural Research Quarterly: JARQ, 48*(3), 355–362.

7. Asyraf, M. R. M., Ishak, M. R., Syamsir, A., Nurazzi, N. M., Sabaruddin, F. A., Shazleen, S. S., Norrrahim, M. N. F., Rafidah, M., Rashid, M. Z. A., & Razman, M. R. (2022). Mechanical properties of oil palm fibre-reinforced polymer composites: A review. *Journal of Materials Research and Technology, 17,* 33–65.
8. Shinoj, S., Visvanathan, R., Panigrahi, S., & Kochubabu, M. (2011). Oil palm fiber (OPF) and its composites: A review. *Industrial Crops and Products, 33*(1), 7–22.
9. Law, K. N., Wan Rosli, W. D., & Ghazali, A. (2007). Oil palm empty fruit bunch, BioResource, 2(3), 351–362.
10. Mardawati, E., Wermen, A., Bley, T., Kresnowati, M., & Sediati, T. (2014). The enzymatic hydrolysis of oil palm EFB to xylose. Journal of Japan Institute of Energy, *93,* 973–978.
11. Rozman, H. D., Saad, M. J., & Mohd Ishak, Z. A. (2003). Flexural and impact properties of oil palm empty fruit bunch (EFB)–polypropylene composites—the effect of maleic anhydride chemical modification of EFB. *Polymer Testing,* 22(3), 335–341.
12. Rao, P. S., Hardiman, M., O'Dowd, N. P., & Sebaey, T. A. (2021). Comparison of progressive damage between thermoset and thermoplastic CFRP composites under in-situ tensile loading. *Journal of Composite Materials, 55,* 1473–1484. https://doi.org/10.1177/0021998320972471
13. Sahari, J., & Maleque, M. A. (2016). Mechanical properties of oil palm shell composites. *International Journal of Polymer Science, 6,* 1–7.
14. Das, S. (2017). Mechanical properties of waste paper/jute fabric reinforced polyester resin matrix hybrid composites. *Carbohydrate Polymers, 172.* https://doi.org/10.1016/j.carbpol.2017.05.036
15. Chaiwong, W., Samoh, N., Eksomtramage, T., & Kaewtatip, K. (2019). Surface-treated oil palm empty fruit bunch fiber improved tensile strength and water resistance of wheat gluten-based bioplastic. *Composites Part B: Engineering, 176,* 107331. https://doi.org/10.1016/j.compositesb.2019.107331
16. Ramlee, N. A., Jawaid, M., Zainudin, E. S., & Yamani, S. A. K. (2019). Modification of oil palm empty fruit bunch and sugarcane bagasse biomass as potential reinforcement for composites panel and thermal insulation materials. *Journal of Bionic Engineering, 16,* 175–188.
17. Ilyas, R. A., Sapuan, M. S., Norizan, M. N., Norrrahim, M. N. F., Ibrahim, R., Atikah, M. S. N., et al. (2020). Macro to nanoscale natural fiber composites for automotive components: Research, development, and application. In M. S. Sapuan & R. A. Ilyas (Eds.), *Biocomposite synthesis, characterization, and application* (pp. 1–50). Woodhead Publishing Series.
18. Ariffin, H., Norrrahim, M. N. F., Yasim-Anuar, T. A. T., Nishida, H., Hassan, M. A., Ibrahim, N. A., et al. (2017). Oil palm biomass cellulose-fabricated polylactic acid composites for packaging applications. In *Bionanocomposites for packaging applications* (pp. 95–105). http://dx.doi.org/10.1007/978-3-319-67319-6_5
19. Hussain, D., Sivakumar, M. A., Daud, M. Z., & Selamat, M. Z. (2016). Tensile performance of palm oil fiber metal laminate. In *Proceedings of mechanical engineering research day,* 121–122.
20. Siyamak, S., Ibrahim, N. A., Abdolmohammadi, S., Yunus, W. M. Z. W., & Rahman, M. Z. A. (2012). Effect of fiber esterification on fundamental properties of oil palm empty fruit bunch fiber/poly(butylene adipate-co-terephthalate) biocomposites. *International Journal of Molecular Sciences, 13*(2), 1327–1346.

21. Rizal, S., Fizree, H. M., Saurabh, C. K., Gopakumar, D. A., Sri Aprilia, N. A., Hermawan, D., et al. (2019). Value-added utilization of agro-waste derived oil palm ash in epoxy composites. *Journal of Renewable Materials, 7*, 1269–1278. https://doi.org/10.32604/jrm.2019.07227
22. Hanan, F., Jawaid, M., Paridah, M. T., & Naveen, J. (2020). Characterization of hybrid oil palm empty fruit bunch/woven kenaf fabric-reinforced epoxy composites. *Polymers, 12*, 9, 2052.
23. Amir, S. M. M., Sultan, M. T. H., Jawaid, M., Safri, S. N. A., Shah, A. U. M., Yusof, M. R., et al. (2019). Effects of layering sequence and gamma radiation on mechanical properties and morphology of Kevlar/oil palm EFB/epoxy hybrid composites. *Journal of Materials Research and Technology, 8*, 5362–5373.
24. Zuhri, M., Yusoff, M., Sapuan, S. M., Ismail, N., & Wirawan, R. (2010). Mechanical properties of short random oil palm fibre reinforced epoxy composites. *Sains Malaysiana, 39*, 87–92.
25. Richard, B. D., Wahi, A., Nani, R., Iling, E., Osman, S., & Ali, D. S. H. (2019). Effect of fiber loading on mechanical properties of oil palm frond/urea formaldehyde (OPF/UF) composite. *International Journal of Integrated Engineering, 11*, 122–128.
26. Sreekala, M. S., Kumaran, M. G., Joseph, S., Jacob, M., & Thomas, S. (2000). Oil palm fibre reinforced phenol formaldehyde composites: Influence of fibre surface modifications on the mechanical performance. *Applied Composite Materials, 7*, 295–329.
27. Malkawi, A. B., Habib, M., Aladwan, J., & Alzubi, Y. (2020). Engineering properties of fibre reinforced lightweight geopolymer concrete using palm oil biowastes. *Australian Journal of Civil Engineering, 18*(1), 82–92. https://doi.org/10.1080/14488353.2020.1721954
28. Pacheco-Torgal, F., Cabeza, L. F., Labrincha, J., & De Magalhaes, A. G. (2014). *Eco-efficient construction and building materials: Life cycle assessment (LCA), ecolabelling and case studies*. Woodhead Publishing.
29. Asdrubali, F., D'Alessandro, F., & Schiavoni, S. (2015). A review of unconventional sustainable building insulation materials. *Sustainable Materials and Technologies, 4*, 1–7.
30. Panyakaew, S., & Fotios, S. (2008). Agricultural waste materials as thermal insulation for dwellings in Thailand: Preliminary results. In *25th Conference on passive and low energy architecture, 321*.
31. Laborel-Préneron, A., Magniont, C., & Aubert, J.-E. (2018). Hygrothermal properties of unfired earth bricks: Effect of barley straw, hemp shiv and corn cob addition. *Energy and Buildings, 178*, 265–278. https://doi.org/10.1016/j.enbuild.2018.08.021
32. Abu-Jdayil, B., Mourad, A. H., Hittini, W., Hassan, M., & Hameedi, S. (2019). Traditional, state-of-the-art and renewable thermal building insulation materials: An overview. *Construction and Building Materials, 214*, 709–735. https://doi.org/10.1016/j.conbuildmat.2019.04.102
33. Ramlee, N. A., Naveen, J., & Jawaid, M. (2021). Potential of oil palm empty fruit bunch (OPEFB) and sugarcane bagasse fibers for thermal insulation application – A review. *Construction and Building Materials, 271*, 121519.

34. Hamzah, M. H., Deraman, R., & Saman, N. S. (2017). Investigating the effectiveness of using agricultural wastes from empty fruit bunch (EFB), coconut fibre (CF) and sugarcane bagasse to produce low thermal conductivity clay bricks. In *AIP conference proceedings* (Vol. 1901, p. 30005). AIP Publishing. https://doi.org/10.1063/1.50104700
35. Hamzah, M. H., Deraman, R., & Saman, N. S. (2017). Investigating the effectiveness of using agricultural wastes from empty fruit bunch (EFB), coconut fibre (CF) and sugarcane bagasse to produce low thermal conductivity clay bricks. In *AIP conference proceedings* (Vol. 1901, p. 30005). AIP Publishing. https://doi.org/10.1063/1.50104700
36. Erwinsyah, E., & Richter, C. (2007). Thermal insulation material made from oil palm empty fruit bunch fibres. *Biotropia, 14*(1), 32–50. https://doi.org/10.11598/btb.2007.14.1.23
37. Hassan, S., Tesfamichael, A., & Nor, M. M. (2014). Comparison study of thermal insulation characteristics from oil palm fibre. In *MATEC web of conferences* (Vol. 13, p. 2016). EDP Sciences. https://doi.org/10.1051/matecconf/20141302016
38. Agrawal, R., Saxena, N. S., Sharma, K. B., Sreekala, M. S., & Thomas, S. (1999). Thermal conductivity and thermal diffusivity of palm fiber reinforced binary phenol-formaldehyde composites. *Indian Journal of Pure and Applied Physics, 37*, 865–869.
39. Malkawi, A. B., Nuruddin, M. F., Fauzi, A., Al-Mattarneh, H., & Mohammed, B. S. (2017). Effect of plasticizers and water on properties of HCFA geopolymers. *Key Engineering Materials, 733*, 76–79. https://doi.org/10.4028/www.scientific.net/KEM.733.76
40. Khalil, H. A., Jawaid, M., Hassan, A., Paridah, M., & Zaidon, A. (2012). *Oil palm biomass fibres and recent advancement in oil palm biomass fibres based hybrid biocomposites.* https://doi.org/10.1094/PDIS-11-11-0999-PDN
41. Malkawi, A. B., Nuruddin, M. F., Fauzi, A., Al-Mattarneh, H., & Mohammed, B. S. (2017). Effect of plasticizers and water on properties of HCFA geopolymers. *Key Engineering Materials, 733*, 76–79. Trans Tech Publications. https://doi.org/10.4028/www.scientific.net/KEM.733.76
42. Nuruddin, M., Malkawi, A., Fauzi, A., Mohammed, B., & Al-Mattarneh, H. (2016). Effects of alkaline solution on the microstructure of HCFA geopolymers. In *Engineering challenges for sustainable future* (pp. 501–505). Routledge in association with GSE Research.
43. Olivia, M., Kamaldi, A., Sitompul, I. R., Diyanto, I., & Saputra, E. (2014). Properties of geopolymer concrete from local fly ash (FA) and palm oil fuel ash (POFA). In *Materials science forum* (Vol. 803, pp. 110–114). Trans Tech Publications, Ltd. https://doi.org/10.4028/www.scientific.net/msf.803.110
44. Neville, A. M. (1995). *Properties of concrete.* Longman Group Limited.
45. Pacheco-Torgal, F., Castro-Gomes, J., & Jalali, S. (2007). Cement and concrete research. *Cement and Concrete Research, 37*, 933
46. Duxson, P., Provis, J. L., Lukey, G. C., Mallicoat, S. W., Kriven, W. M., & van Deventer, J. S. J. (2006). Colloids and surfaces A: Physicochemical and engineering aspects. *Colloids and Surfaces A: Physicochemical and Engineering Aspects, 269*, 47.

47. Mohajerani, A., Ukwatta, A., Jeffrey-Bailey, T., Swaney, M., Ahmed, M., Rodwell, G., Bartolo, S., Eshtiaghi, N., & Setunge, S. (2019). A proposal for recycling the world's unused stockpiles of treated wastewater sludge (bio-solids) in fired-clay bricks. *Buildings, 9*(1), 10. https://doi.org/10.3390/buildings9010010
48. Kazmi, S. M. S., Munir, M. J., Patnaikuni, I., Wu, Y. F., & Fawad, U. (2018). Thermal performance enhancement of eco-friendly bricks incorporating agro-wastes. *Energy and Buildings, 158,* 1117–1129. https://doi.org/10.1016/j.enbuild.2017.10.056
49. Sata, V., Jaturapitakkul, C., & Kiattikomol, K. (2004). Utilization of palm oil fuel ash in high-strength concrete. *Journal of Materials in Civil Engineering, 16*(6), 623–628.
50. Tjaronge, M. W., & Caronge, M. A. (2021). Physico-mechanical and thermal performances of eco-friendly fired clay bricks incorporating palm oil fuel ash. *Materialia, 17,* 101130.
51. Kazmi, S. M. S., Abbas, S., Saleem, M. A., Munir, M. J., & Khitab, A. (2016). Manufacturing of sustainable clay bricks: Utilization of waste sugarcane bagasse and rice husk ashes. *Construction and Building Materials, 120,* 29–41.
52. Sari, Y. W., Syafitri, U., Sanders, J. P. M., & Bruins, M. E. (2015). How biomass composition determines protein extractability. *Industrial Crops and Products, 70,* 125–133.
53. Sari, Y. W., Silviana, M. M., Kurniati, M., & Budiman, I. (2017). Valorization of palm kernel cake as bioadhesive for particle board. *IOP Conference Series: Earth and Environmental Science, 187,* 1–10.
54. Hee, C. W., Shing, W. L., & Wei, L. X. (2020). Optimization treatment of palm kernel cake for the production of bioadhesive for plywood. *INTI Journal, 2020,* 19.
55. Soi, H. S., Sattar, M. N., Tuan Ismail, T. N. M., Palam, K. D. P., Hamzah, N. A., Kian, Y. S., Hassan, H. A., & Ahmad, S. (2009). *Palm oil-based adhesives for fibre board* (MPOB Information Series, MPOB TT no. 441). http://tot.mpob.gov.my/tt-no-441-palm-oil-based-adhesive-for-fibre-board/
56. Kurniawati, A. S., Pratiwi, A. I., Ariyani, N. R., Septiani, M., & Sasongko, N. A. (2022). Utilization of palm oil based biofuel to support sustainable energy transition in Indonesia. *IOP Conference Series: Earth and Environmental Science, 1108,* 012038.
57. Pramanik, A., Chaudhary, A. A., Sinha, A., Chaubey, K. K., Ashraf, M. S., Basher, N. S., Rudayni, H. A., Dayal, D., & Kumar, S. (2023). Nanocatalyst-based biofuel generation: An update, challenges and future possibilities. *Sustainability, 15*(7), 6180. https://doi.org/10.3390/su15076180
58. Roundtable on Environmental Health Sciences, Research, and Medicine; Board on Population Health and Public Health Practice; Institute of Medicine. (2014). *The nexus of biofuels, climate change, and human health: Workshop summary.* National Academies Press. https://www.ncbi.nlm.nih.gov/books/NBK196448/
59. El Araby, R., Amin, A., El Morsi, A. K., & El-Ibiari, N. N. (2017). Study on the characteristics of palm oil–biodiesel–diesel fuel blend. *Egyptian Journal of Petroleum, 27*(2), 187–194.

8

Coconut Shell: Solution for Environmental Sustainability

8.1 Introduction

Coconut (*Cocos nucifera L.*) is a very versatile and multifunctional plant that grows abundantly in tropical areas. Indonesia, Malaysia, and India are the largest-producing countries in the world. Almost all parts of the coconut plant have high economic and ecological value. Coconuts produce fibre for rope and planting media, shells which can be converted into charcoal for fuel and activated carbon, pulp for coconut milk and oil, and coconut water for rehydration drinks. The stems are used as construction materials, the leaves for roofing and crafts, while the roots have traditional medicinal value. This diverse use not only provides economic benefits but also supports environmental sustainability by minimizing waste and increasing the added value of coconut products [1].

In line with the growing need for engineering materials, coconut shells can be utilized in various applications [2, 3]. Coconut shells have the potential to be used as raw materials in the construction industry, as additional ingredients in lightweight concrete mixtures, or as a substitute for conventional building materials. Apart from that, coconut shells can also be processed into activated charcoal, which is used in water treatment and pollutant removal. The use of coconut shells in various applications not only utilizes agricultural waste effectively but also supports the principle of sustainability by reducing the use of synthetic materials and reducing environmental impacts. Coconut shell powder is also widely used in the plywood and laminated board industry, as well as in synthetic resin glue, mosquito coils, and incense sticks. Its consistent quality and chemical composition make it a superior alternative to bark powder, furfural, and peanut shell powder.

DOI: 10.1201/9781003617594-8

8.2 Coconut Shell Reinforced Composites

Composites are increasingly in demand to meet the optimal criteria for substituting metal materials. Advances in polymer technology have facilitated the development of lightweight and strong materials. Researchers are now concentrating on exploring various combinations of biodegradable matrices and natural fillers to create biodegradable composite materials that provide enhanced mechanical properties while simultaneously reducing production costs. **Figure 8.1** illustrates the various parts of the coconut shell.

Compared with traditional reinforcing fibres such as glass and carbon, natural fillers provide advantages such as more affordable cost, high strength, corrosion resistance, low density, and excellent specific strength properties. The use of natural fibres in the automotive industry includes being able to reduce weight by 10 to 30% with mechanical properties that are still acceptable [4]. However, natural composites also face challenges such as

FIGURE 8.1
Coconut tree, coconut fruit, and coconut shell (Illustration: canva, 2024, https://www.canva.com).

TABLE 8.1
Characteristics of Coconut Shells [5]

No	Physical and Mechanical Properties	Coconut Shells
1	Moisture content (%)	4.20
2	Water absorption (24 h) (%)	24.00
3	Specific gravity	1.05–1.20
4	SSD	1.40–1.50
5	Impact value (%)	8.15
6	Crushing value (%)	2.58
7	Abrasion value (%)	1.63
8	Bulk density (kg/m^3)	650
9	Compacted loose	550
10	Fineness modulus	6.26
11	Shell thickness (mm)	2–8

lower tensile strength, lower melting point, unsuitability for high-temperature applications, poor adhesion to hydrophobic polymers, varying particle sizes, and susceptibility to moisture-induced degradation. Therefore, chemical treatments are used to modify the surface properties of the fibres and overcome these limitations. **Table 8.1** shows the physical and mechanical characteristics of coconut shells [5].

Coconut shell, as an alternative composite filler, is a strong and lightweight material that is abundantly available as agricultural waste. In engineering applications, coconut shells are frequently utilized as fillers or reinforcements in various composite materials, including epoxy resin and other polymers. The natural properties of coconut shells, combined with a polymer matrix, yield composites with excellent tensile strength, wear resistance, and extensive application potential across various industries. The replacement of synthetic and natural fibres with coconut shells demonstrates significant promise, with mechanical properties influenced by the type of matrix, reinforcement ratio, and treatment methods or chemical compositions used. Therefore, investigating the effects of coconut shell fibres in composites involving different matrices and treatments is essential for expanding their applications.

Numerous studies have explored the use of coconut shell fibres. For instance, composites made with epoxy resin (MGSL 285) using the hand lay-up technique exhibited a tensile strength of 28 MPa and a hardness of 77 [6]. In another study, recycled polypropylene treated with 1 molar *sodium hydroxide* (NaOH) solution displayed a tensile strength of 11 N/mm^2, a hardness of 12 HBR, an impact energy of 11.5 J, and a water absorption rate of 8% [7]. Furthermore, pretreating coconut shells with *sodium hydroxide* (NaOH) and *potassium hydroxide* (KOH) and combining them with epoxy resin as a matrix produced a flexural strength of 32.54 MPa and an impact strength of 42.57 MPa [8]. Composites using epoxy resin (*Bisphenol-A-Co-Epichlorohydrine*) and coconut shell as a natural fibre achieved a tensile strength of 20 N/mm² and a hardness of 61 DHN [9]. *Rahul* (2012) [10] utilized unsaturated polyester (USP) resin and coconut shell fibres treated with 1% (v/v) sodium hydroxide (NaOH) through compression moulding techniques, resulting in a tensile strength of 33 MPa, a breaking strength of 4%, a flexural strength of 57 MPa, an elastic modulus of 1500 MPa, a flexural modulus of 2800 MPa, and thermal stability up to 600°C in the composite matrix. Additionally, *Sarena* (2012) [11] employed natural rubber as a matrix, treating it with 10 mL of NaOH solution for 5 hours, which yielded a tensile strength of 25 MPa, Young's modulus of 4.5 MPa, a fracture strength of 5.6 N/mm², a hardness of 50 (Shore A), and thermal stability up to 90°C. The relationship between coconut shell filler and the mechanical properties of the matrix indicates that composites with 25% and 50% filler exhibit superior hardness, tensile strength, elongation at the breaking point, and specific gravity compared to other combinations, with 50% filler composites demonstrating greater effectiveness than those with 25% filler [12].

To improve the mechanical properties of coconut shell composites, it can also be done with other natural materials. *Somashekhar T*, 2018 [13], conducted experimental tests on the mechanical properties of composites reinforced with coconut shell powder and tamarind seed powder. Three different percentages of coconut shell powder and epoxy resin were used to form the composite material, and the results were analyzed for these variations. The findings showed that the addition of tamarind seed powder with coconut shell powder significantly improved the tensile properties, with an increase of approximately 50%. Optimal mechanical properties were achieved with a composition of 50% coconut shell powder, 5% tamarind seed powder, and 45% epoxy resin. Experimental results on changes in the mechanical properties of various composite samples filled with coconut shell powder (CSP) have been reported [14]. Composite samples containing 20% coconut shell powder (CSP) exhibited a flexural strength of 91.25 MPa and a tensile strength of 21.55 MPa. Increasing the CSP content to 30% resulted in an enhanced flexural strength of 94.55 MPa, although the tensile strength decreased to 19.21 MPa. However, a further increase in CSP content to 40% led to a decline in flexural strength, which was recorded at 78.95 MPa, along with a tensile strength of 16.45 MPa. These results indicate that there is an optimal CSP content that maximizes the mechanical properties. Exceeding this optimal level can negatively affect the material's mechanical performance.

8.2.1 Geopolymer Concrete

One of the efforts to produce environmentally friendly concrete has led to the development of concrete using inorganic binders, especially *alumina-silicate* polymers known as geopolymers [15]. Geopolymers can be made from natural geological materials or industrial by-products such as fly ash, which is rich in silica and alumina. The use of geopolymers in concrete not only reduces carbon emissions associated with conventional Portland cement production but also utilizes industrial waste, thereby reducing the overall environmental impact. With technological advances, geopolymer concrete can be designed to have superior mechanical properties such as high compressive strength and good resistance to chemical attack, making it a promising alternative for sustainable construction.

Fly ash is an industrial byproduct generated from burning unused coal and is utilized in the production of geopolymer concrete binders. This material is very fine and possesses a high cement content along with pozzolanic properties, enabling it to react with free lime in cement during hydration to form a binding compound in the presence of water at normal temperatures. Fly ash is activated using alkaline solutions, such as sodium hydroxide (NaOH) and sodium silicate (Na_2SiO_3), which serve as catalysts and play a crucial role in the geopolymerization process [16].

Currently, researchers are investigating the potential of using coconut shell ash (ATK) as a cement substitute in environmentally friendly concrete [16–18]. Coconut shells are an inevitable byproduct of the agricultural sector, and ATK is produced through the combustion of these shells. Studies show that using ATK as a cement substitute in quantities ranging from 0 to 20% of the cement weight can lead to maximum compressive strength when 12.5% ATK is used. Tests conducted at various temperatures (100°C, 200°C, and 300°C) for one hour indicate that concrete containing ATK demonstrates better strength and weight reduction compared to ordinary concrete [17].

Additionally, experiments with reinforced concrete beams incorporating 10% ATK and 5% coconut shell coarse aggregate revealed enhancements in flexural and shear capacities, alongside a 17.3% reduction in bending load, which indicates increased ductility. This study confirmed that the inclusion of ATK and coconut shell particles improves ductility while also moderating the pozzolanic reaction of ATK [19, 20]. ATK is characterized by high alumina content and an irregular, elongated shape compared to other agricultural waste ashes, which contributes to a lower slump value in concrete mixtures [20, 21]. Overall, concrete mixes with a 10% ATK replacement exhibit superior strength properties, reduced water absorption rates, and enhanced durability compared to other composite concretes.

8.2.2 Composite Concrete with Coconut Shell

The increasing demand for concrete structural buildings has increased the need for coarse and fine aggregates. Annually, approximately 7.23 billion tons of concrete are consumed [22], with aggregates accounting for 60–70% of concrete preparation [23]. As a result, high aggregate consumption has increased costs and depleting resources. To reduce environmental impacts, coconut shells (CS) can be used as a substitute for coarse aggregate (CA). CS, as an abundant agricultural waste, offers a natural alternative to obtain economical and environmentally friendly concrete [24] while saving energy and being recyclable [25]. The properties of coconut shell (CS), which has a specific gravity of 1.24, water absorption of 2.12%, fineness modulus of 5.12, and density of 648 kg/m³, are suitable for use as a partial replacement in concrete mixtures [26]. Currently, researchers [25, 27, 28] worldwide are exploring alternatives to coarse aggregate (CA) in concrete production. A study conducted by Ranjitha in 2021 [26] evaluated the use of concrete incorporating varying percentages of coconut shell (CS)—specifically 0%, 5%, 10%, 15%, 20%, 25%, and 30%—to achieve optimal strength and lightweight properties. The findings indicated that CS can effectively replace up to 20% of coarse aggregate without compromising the concrete's strength and rheological properties.

Additionally, research [5] has shown that concrete with coconut shell filler (CSC) exhibits improved workability due to its smooth surface. The density of CSC ranges from 1930 to 1970 kg/m³, classifying it as lightweight concrete (less than 2000 kg/m³). The flexural strength of CSC is approximately 17.53% and 16.42% of the compressive strength, with values of 26.70 N/mm² and 25.95 N/mm², respectively. Its tensile strength is about 10.11% and 9.17% of the compressive strength. Notably, CSC demonstrates higher impact resistance compared to conventional concrete and meets the criteria outlined in IS 456:2000 and BS 8110. Overall, CSC fulfils the necessary requirements for use as lightweight concrete.

Research on the comparison of costs and strength characteristics of concrete using crushed, granular, and original coconut shells was carried out by *Olanipekun*, 2006 [29]. They used palm oil in different conditions where the composition ratio was between crushed shell, granular, and kernel as a substitute for conventional coarse aggregate. The results show that the compressive strength decreases with increasing shell content in both mixture ratios (1:1:2 and 1:2:4). However, concrete with coconut shells has a higher compressive strength than palm kernel shell concrete. In addition, cost reductions of 30% for coconut shells and 42% for palm kernel shells. This study concluded that coconut shells are more suitable than palm kernel shells as a substitute for conventional aggregates based on the strength/economic ratio.

8.3 Biochar

Coconut shell charcoal (biochar) is made by *pyrolysis* where the coconut shell is heated at a high temperature without involving oxygen in the combustion so that the decomposition of the material occurs. The *pyrolysis* process is carried out under various temperature conditions of 300°C–500°C, for 15 minutes - 60 minutes, and particle sizes of 5.0 mm–25.0 mm [30]. During this process, the coconut shell loses about 70% of its original weight. Analysis of the composition of coconut shell charcoal at a temperature of 900°C shows

TABLE 8.2

Composition Charcoal Shell Chemistry Coconut [32]

Composition	Content (%)
K_2O	45.01
Na_2O	15.42
CaO	6.26
MgO	1.32
Fe_2O_3 and Al_2O_3	1.39
P_2O_3	4.64
SO_3	5.75
SiO_2	4.64

the following parameters: ash 1.64%, carbon 99.27%, hydrogen 0.16%, nitrogen 0.62%, total sulphur 0.05% [31]. Apart from that, there are other chemical compositions of coconut shell charcoal (**Table 8.2**). High-quality coconut shell charcoal comes from coconut shells that are dry and fully ripe. It is an excellent raw material for making high-quality activated carbon. Coconut shell charcoal is widely used for heating and cooking purposes because of its environmentally friendly nature and clean burning.

The quality of charcoal is indicated by the heating value (HV), which depends on the carbonization temperature and additional materials such as coal. The research results indicate that the calorific value of charcoal fuel increases with temperature, rising from 300°C to 500°C, across varying coal masses of 30, 20, 15, and 10 g [33]. This increase in calorific value is attributed to the carbonization process, which enhances the carbon content in biomass and consequently improves fuel efficiency [34]. However, when the coal mass is set at 25 g, a temperature change from 350°C to 400°C results in a decrease in heating value [35]. This decline may be due to interactions between independent variables and fixed factors, such as the amount of adhesive used or moulding pressure, which were not accounted for or assumed to have minimal impact in this study.

Coconut shell biochar presents a versatile and environmentally friendly solution across various applications. Known for its high porosity and substantial carbon content, coconut shell biochar is suitable for uses ranging from organic fertilizers in agriculture to environmental remediation efforts. Biomass charcoal derived from coconut shells can be transformed into organic fertilizer, offering multiple benefits (**Figure 8.2**). Notably, biomass charcoal is effective in reducing antibiotic resistance genes (ARGs) in the soil environment, demonstrating a reduction of 61.54% in ARGs. This biochar exhibits strong adsorption capabilities, with a large specific surface area of 476.66 m^2/g and pore characteristics (diameter approximately 1.207 nm, with a total pore volume of 0.2451 m^3/g) [36]. These properties enhance soil heterogeneity, creating a barrier that limits the proliferation of resistant bacteria and the spread of ARGs. Previous studies have also utilized coconut shell

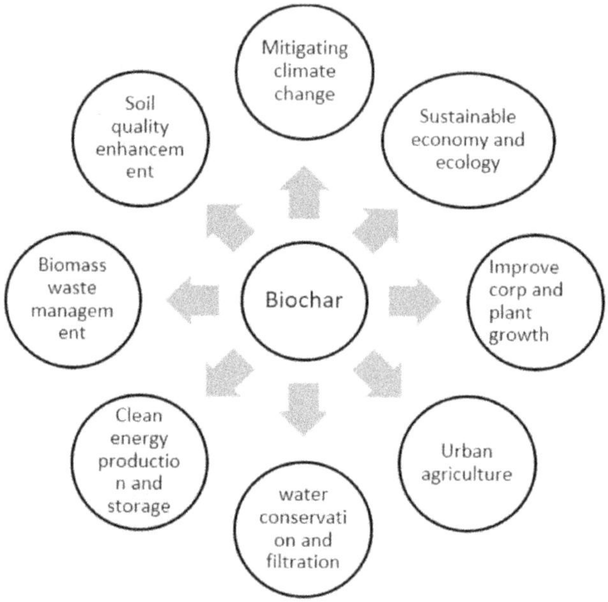

FIGURE 8.2
Various uses of charcoal [38].

biochar to remove contaminants, such as phenols (*Hao et al.*, 2018) [37] and heavy metals (*Paranavithana et al.*, 2016), from the soil environment [36].

Coconut shell charcoal can be manufactured in various types and sizes to meet various needs and applications. This versatility allows it to be used for purposes such as shisha, grilling, or industrial applications. By selecting the appropriate size and shape, users can optimize performance, including burning time and heat output, ensuring the charcoal meets their specific requirements. Additionally, choosing a trusted and professional manufacturer is crucial for ensuring the quality and consistency of coconut shell charcoal products.

8.4 Traditional Process of Charcoal Production

The traditional process of charcoal production involves several stages. First, wood is obtained from forests or community plantations [39]. Once harvested, the wood is stored for at least 120 days to reduce humidity and water content, then cut into the desired size of between 1 m and 1.4 m. The wood is put into a kiln (furnace) where the temperature and oxygen content can be controlled. The next stage is ignition, with all intake and exhaust ports remaining open to control oxygen and heating. This stage aims to remove the water content in the wood and reduce humidity by regulating oxygen input to maintain the ideal temperature. Next is the carbonization stage. At this stage the temperature is reduced gradually, with the main aim of forming carbon and producing charcoal. In this stage, the heating rate, pressure, and final temperature are very important to pay attention to. Oxygen levels are controlled empirically by monitoring the colour of the smoke. After reaching the final temperature, the kiln is closed, and natural cooling occurs until the temperature drops below 40°C [40]. Finally, the charcoal is removed, packaged, and shipped, with increased residue production due to low density and brittleness.

8.4.1 Types of Furnaces

Furnaces are used to make biochar. Traditional stoves, often found in rural industries, are usually made of brick or clay (**Figure 8.3**). There are four types of combustion furnaces used in the biochar manufacturing process, each with different characteristics and operational methods, namely [41]:

- Brazilian beehive kiln

The combustion process in the furnace is regulated to decrease gradually. The operator uses the colour of the smoke emitted from the furnace as a

FIGURE 8.3
Traditional charcoal furnace (Illustration: bing AI, 2024, https://www.bing.com/images/create).

signal. When blue smoke is visible, they close the holes to direct the fire downward. Once the blue smoke is seen coming from the lower ventilation, the furnace is sealed and covered with a mixture of mud to assist in the cooling process.

- Modified Brazilian beehive kiln

This furnace does not have its own ignition hole, so ignition is done through the upper door. After approximately three hours, the hole is closed to distribute the fire. The fire control process is similar to that used in the traditional Brazilian Beehive Furnace.

- Argentinian half-orange kiln

The furnace is lit from the top entrance. After approximately three hours, the hole is closed.

- Metallic industrial type kiln

This system involves four connected furnaces, all of which must be loaded to start the production process. The furnaces are ignited with an external fire, with wood burnt beneath each furnace until an exothermic process is achieved (>280°C). The firewood must have the lowest possible moisture content for successful ignition.

Making charcoal from coconut shells begins with the preparation stage, which includes providing wood, drying, and special treatments to ensure the charcoal has the desired characteristics, as shown schematically in **Figure 8.4**.

- Carbonization process in a furnace

The carbonization process begins with drying the wood in a furnace at a temperature of no more than 100°C until it reaches zero moisture content [42]. Next, the temperature is raised to around 280°C, with the energy coming from the partial combustion of the wood itself, which is an endothermic reaction. At this stage, the wood begins to decompose spontaneously, releasing water vapour, *methanol, acetic acid, tar,* and gases such as hydrogen, carbon monoxide, and carbon dioxide. Air is introduced into the furnace to allow combustion to occur, which also introduces nitrogen into the gas mixture. When temperatures exceed 280°C, wood decomposition becomes exothermic, releasing energy until only charcoal remains. Without additional heating, the process stabilizes at around 400°C. The resulting charcoal contains residual tar (around 30% by weight), ash (3–5%), and the remainder is bound carbon (65–70%). Further heating can reduce the tar content, increasing the bound carbon to about 85% at 500°C, with the volatile content

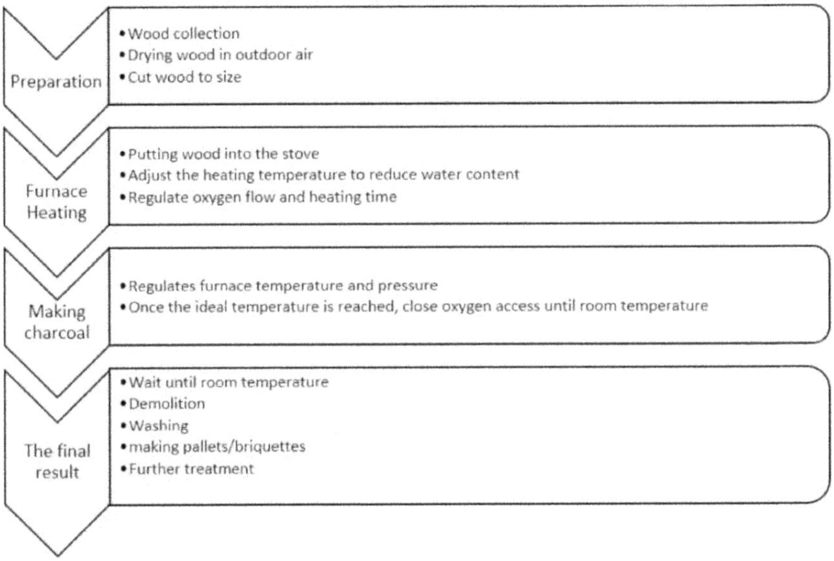

FIGURE 8.4
Schematic of making charcoal from coconut shells [39].

reduced to about 10%. The char yield, approximately 33% of the initial oven-dry wood weight, varies with carbonization temperature due to changes in volatile matter content [43–45]. The charcoal yield depends on temperature warmup as shown in **Table 8.3**.

Lower temperatures during carbonization result in a greater amount of charcoal being produced, but the quality tends to be lower. Charcoal produced at lower temperatures is often of lower quality, contains acidic tar which can be corrosive, does not burn cleanly, and emits smoke. High-quality commercial charcoal typically requires a final carbonization temperature of around 500°C to achieve a fixed carbon content of around 75% [43–45].

TABLE 8.3
The Effect of Temperature on Results Charcoal Composition in the Carbonization Process [43–45]

Carbonization Temperature	Chemical Analysis of Charcoal		Charcoal Yield Based on Oven-Dry Wood
°C	% of fixed charcoal	% volatile materials	(0% moisture)
300	68	31	42
500	86	13	33
700	92	7	30

8.4.2 Charcoal Characterization

- Charcoal

Charcoal is the product obtained from a specific amount of biomass during the pyrolysis process [43]. Charcoal is typically expressed in terms of volume or mass produced per unit mass of the original biomass material. The yield of charcoal is influenced by various factors, including the type of biomass, pyrolysis conditions (such as temperature, heating rate, and residence time), and the production method. These parameters are crucial for evaluating the efficiency and economic feasibility of charcoal production processes. Additionally, the quality of the charcoal is also dependent on the material used for its production.

Time is a critical factor that affects the carbonization process. During this process, the temperature is gradually increased to remove water and volatile content from the biomass, resulting in the production of charcoal. The relationship between the heating process, heating duration, and biomass mass loss is illustrated in **Figure 8.5**. As heating progresses, biomass undergoes thermal decomposition, leading to a decrease in mass as the temperature rises. This process begins with the evaporation of water at lower temperatures, followed by the decomposition of volatile compounds at higher temperatures. During this stage, the chemical components in the biomass, such as cellulose, hemicellulose, and lignin, break down into gases and tar, leaving behind solid carbon.

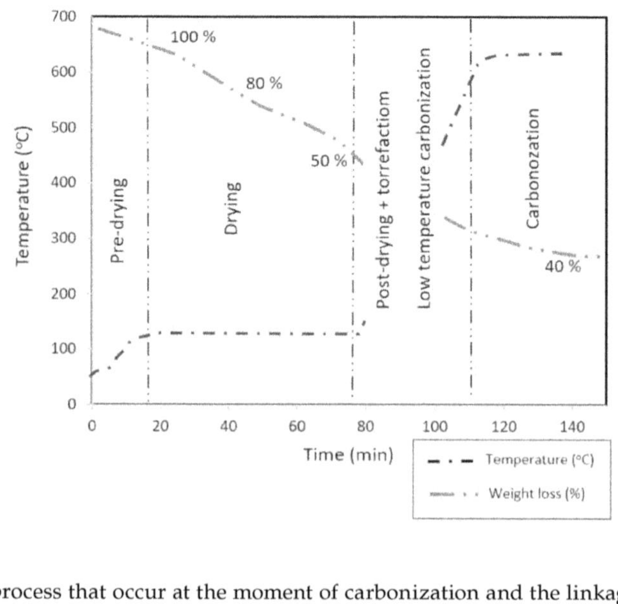

FIGURE 8.5
Stages of the process that occur at the moment of carbonization and the linkage between the heating process and time warmup to loss of biomass mass [46].

- Physicochemical analysis

The physicochemical analysis of charcoal encompasses the physical and chemical characteristics that define its properties. This analysis determines key attributes such as water content, volatile matter, fixed carbon, and ash content. The heating process, including temperature and duration, significantly impacts the composition of charcoal, affecting parameters such as moisture content (MC), volatile matter (VM), ash content, and fixed carbon (FC) (see **Figure 8.6**). Below is a brief explanation of each of these general parameters:

1. **Water content (MC)**: the amount of water contained in charcoal which affects efficiency, combustion, and value energy. High water content can reduce the heat capacity of charcoal.
2. **Volatile compounds (VC)**: this parameter refers to the proportion of substances that easily vapourize when charcoal is heated to a specific temperature. These include organic compounds that can significantly influence the combustion characteristics and ignition properties of the charcoal.
3. **Fixed carbon (FC)**: this refers to the solid residue remaining after the volatile compounds have evaporated and the moisture has been removed. Fixed carbon is an important measure of the carbon content available for combustion and directly influences the heating value of the charcoal.

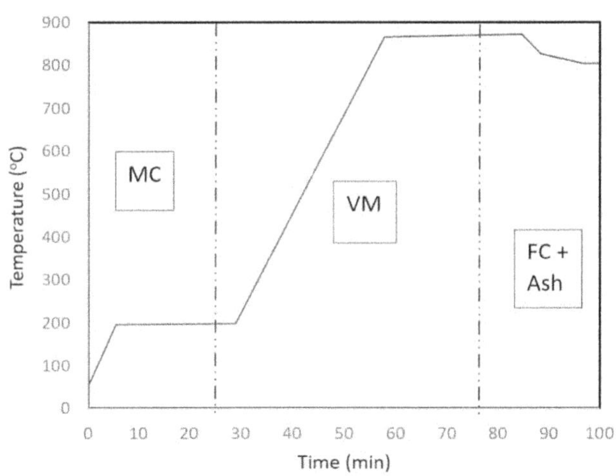

FIGURE 8.6
Relationship between temperature and heating time on the composition of charcoal. Note: *moisture content (MC), volatile matter (VM), ash (ASH),* and *fixed carbon (FC)* [46, 47].

4. **Ash content**: this is the inorganic residue that remains after combustion, primarily composed of minerals present in the original biomass. A high ash content can negatively affect the handling and burning properties of charcoal.

8.4.3 Calorific Value

The high calorific value of charcoal, an important indicator of its energy content, is precisely measured using an automatic iso-peribolic calorimeter (*Leco's AC600*), following strict standards as outlined in UNE-EN 18125 [48]. The calorific value, or heating value, represents the energy released when the material, such as charcoal, is completely combusted under specific conditions. This value indicates the potential energy that can be produced from burning the material. For charcoal, a higher heating value signifies its ability to generate a significant amount of heat when utilized as a fuel source. The empirical heat produced is influenced by the composition of elements such as nitrogen, carbon, oxygen, hydrogen, and sulphur, as formulated in **Equation 1** [49]:

$$\text{HHV (MJkg)} = -4.9140 + 0.2611\,[\%\text{N}] + 0.4114\,[\%\text{C}] + 0.6114\,[\%\text{H}] - 0.02097\,[\%\text{O}] + 0.3888\,[\%\,\text{S}] \quad (1)$$

Where %N, %C, %H, %O, and %S are the mass percentages of the elements nitrogen, carbon, hydrogen, oxygen, and sulphur based on *dry ash-free ultimate analysis*.

References

1. Henrietta, H. M., Kalaiyarasi, K., & Raj, A. S. (2022). Coconut tree (Cocos nucifera) products: A review of global cultivation and its benefits. *Journal of Sustainability and Environmental Management, 1*(2), 257–264. https://doi.org/10.3126/josem.v1i2.45377
2. Pradhan, S. K., Dwarakadasa, E. S., & Reucroft, P. J. (2004). Processing and characterization of coconut shell powder filled UHMWPE. *Material Science and Engineering, 367*, 57–62.
3. Sarki, J., Hassan, S. B., Aigbodion, V. S., & Oghenevweta, J. E. (2011). Potential of using coconut shell particle fillers in eco-composite materials. *Journal of Alloys and Compounds, 509*(5), 2381–2385.
4. Udhayasankar, R., et al. (2015). A review on coconut shell reinforced composites. *International Journal of ChemTech Research, 8*(11), 624–637.
5. Gunasekaran, K., Kumar, P. S., & Lakshmipathy, M. (2011). Mechanical and bond properties of coconut shell concrete. *Construction and Building Materials, 25*(1), 92–98.

6. Ozsoy, N., Ozsoy, M., & Mimaroglu, A. (2014). Comparison of mechanical characteristics of chopped bamboo and chopped coconut shell reinforced epoxy matrix composite materials. *European International Journal of Science and Technology, 3*(8), 15–20.
7. Agunsoye, J. O., Bello, S. A., Azeez, S. O., Yekinni, A. A., & Adeyemo, R. G. (2014). Recycled polypropylene reinforced coconut shell composite: Surface treatment morphological, mechanical and thermal studies. *International Journal of Composite Materials, 4*(3), 168–178.
8. Vignesh, K., Sivakumar, K., Prakash, M., Palanivel, A., & Sriram, A. (2015). Experimental analysis of mechanical properties of alkaline treated coconut shell powder-polymer matrix composites. *International Journal of Advances in Engineering*, 448–453.
9. Akindapo, J. O., Harrison, A., & Sanusi, O. M. (2014). Evaluation of mechanical properties of coconut shell fibers as reinforcement material in epoxy matrix. *International Journal of Engineering Research & Technology, 3*, 2337–2348.
10. Chanap, R. (2012). *Study of mechanical and flexural properties of coconut shell ash reinforced epoxy composites* (BTech thesis). Department of Mechanical Engineering, National Institute of Technology.
11. Sareena, C., Ramesan, M. T., & Purushothaman, E. (2012). Utilization of coconut shell powder as a novel filler in natural rubber. *Journal of Reinforced Plastics and Composites, 31. 8*, 533–547.
12. Keerthika, B., Umayavalli, M., Jeyalalitha, T., & Krishnaveni, N. (2016). Coconut shell powder as cost effective filler in copolymer of acrylonitrile and butadiene rubber. *Ecotoxicology and Environmental Safety, 130*, 1–3.
13. Somashekhar, T. M., Naik, P., Vighnesha, Nayak, M., & Rahul, S. (2018). Study of mechanical properties of coconut shell powder and tamarind shell powder reinforced with epoxy composites. *IOP Conference Series: Materials Science and Engineering, 376*, 012105.
14. Singh, A. (2014). Characterization of novel coconut shell powder reinforced-epoxy composite. *Journal of Engineering and Technology Research, 81, 7.*
15. Zulfiati, R., et al. (2020). The nature of coconut fibre fly ash-based mechanical geopolymer. *IOP Conference Series: Materials Science and Engineering, 807*, 01204.
16. Hardjito, D., Wallah, S. E., & Rangan, B. V. (2004). Factors influencing the compressive strength of fly ash-based geopolymer concrete. *Civil Engineering Dimension, 6*(2), 88.
17. Priya, S. S., Shanmuga, P., & Padmanaban, I. (2024). Effect of coconut shell ash as an additive on the properties of green concrete. *Global NEST Journal, 26*(1). https://doi.org/10.30955/gnj.005413.
18. Joshua, O. (2018). Data on the pozzolanic activity in coconut shell ash (CSA) for use in sustainable construction. *Data in Brief, 18*, 1142–1145. https://doi.org/10.1016/j.dib.2018.03.125.
19. Herring, T. C., Nyomboi, T., & Thuo, J. N. (2022). Ductility and cracking behavior of reinforced coconut shell concrete beams incorporated with coconut shell ash. *Results in Engineering, 14*. https://doi.org/10.1016/j.rineng.2022.100401.
20. Herring, T. C., Thuo, J. N., & Nyomboi, T. (2022). Engineering and durability properties of modified coconut shell concrete. *Civil Engineering Journal, 8*(2), 362–381.

21. Charitha, V., Athira, V. S., Jittin, V., Bahurudeen, A., & Nanthagopalan, P. (2021). Use of different agro-waste ashes in concrete for effective upcycling of locally available resources. *Construction and Building Materials, 285*, 122851. https://doi.org/10.1016/j.conbuildmat.2021.122851.
22. Manjunatha, M., Raju, K. V. B., & Sivapullaiah, P. V. (2021). Effect of PVC dust on the performance of cement concrete—A sustainable approach. In *Lecture notes in civil engineering* (pp. 607–617). Springer. https://doi.org/10.1007/978-981-15-4577-1_52
23. Shruthi, V. A., Tangadagi, R. B., Shwetha, K. G., Nagendra, R., Ranganath, C., Ganesh, B., & Mahesh Kumar, C. L. (2021). Strength and drying shrinkage of high strength self-consolidating concrete. In *Lecture notes in civil engineering* (pp. 615–624). Springer Science and Business Media Deutschland GmbH. https://doi.org/10.1007/978-981-15-5195-6_48
24. Bhuvaneshwari, S., & Ravi, A. (2020). Development of sustainable green repair material using fibre reinforced geopolymer composites. *Journal of Green Engineering, 10*, 494–510.
25. Colangelo, F., Petrillo, A., Cioffi, R., Borrelli, C., & Forcina, A. (2018). Life cycle assessment of recycled concretes: A case study in southern Italy. *Science of the Total Environment, 615*, 1506–1517. https://doi.org/10.1016/j.scitotenv.2017.09.107
26. Tangadagi, R. B., Manjunatha, M., Bharath, A., & Preethi, S. (2020). Utilization of steel slag as an eco-friendly material in concrete for construction. *Journal of Green Engineering (JGE), 10*, 2408–2419.
27. Shah, M. C., Gupta, K. K., Nainwal, A., Negi, A., & Kumar, V. (2021). Investigation of mechanical properties of concrete with natural aggregates partially replaced by recycled coarse aggregate (RCA). *Materials Today: Proceedings*. https://doi.org/10.1016/j.matpr.2020.12.456
28. Tangadagi, R. B., Manjunatha, M., Preethi, S., Bharath, A., & Reshma, T. V. (2021). Strength characteristics of concrete using coconut shell as a coarse aggregate—a sustainable approach. *Materials Today: Proceedings, 47*(Part 13), 3845–3851.
29. Olanipekun, E. A., Olusola, K. O., & Ata, O. (2006). A comparative study of concrete properties using coconut shell and palm kernel shell as coarse aggregates. *Building and Environment, 41*(3), 297–301.
30. Ahmad, R. K., et al. (2020). The effects of temperature, residence time, and particle size on a charcoal produced from coconut shell. *IOP Conference Series: Materials Science and Engineering, 863*, 012005.
31. Pramono, A. E., Zulfia, A., & Soedarsono, J. W. (2011). Physical and mechanical properties of carbon-carbon composite based on coconut shell waste. *Journal of Materials Science and Engineering, 5*, 12–19.
32. Satriawan, A., Yuliet, R., & Permana, D. (2021). Utilization of coconut shell charcoal to improve bearing capacity of clay as subgrade for road pavement. *IOP Conference Series: Earth and Environmental Science*, 832(1), 012041.
33. Li, H., Wang, X., Tan, L., Li, Q., Zhang, C., Wei, X., Wang, Q., Zheng, X., & Xu, Y. (2022). Coconut shell and its biochar as fertilizer amendment applied with organic fertilizer: Efficacy and course of actions on eliminating antibiotic resistance genes in agricultural soil. *Journal of Hazardous Materials, 437*, 129322.
34. Pestaño, L. D. B., & Jose, W. I. (2016). *International Journal of Renewable Energy Development, 5*, 3. 187.

35. Siswanto, Nurma Wahyusi, K., & Panjaitan, R. (2021). Charcoal fuel from the mixture of coconut shell waste and coal: Effect of carbonization temperature and the amount of coal mass in the mixture. *E3S Web of Conferences, 328,* 01019. ICST 2021.
36. Li, H., Wang, X., Tan, L., Li, Q., Zhang, C., Wei, X., Wang, Q., Zheng, X., & Xu, Y. (2022). Coconut shell and its biochar as fertilizer amendment applied with organic fertilizer: Efficacy and course of actions on eliminating antibiotic resistance genes in agricultural soil. *Journal of Hazardous Materials, 437,* 129322.
37. Hao, Z., Wang, C., Yan, Z., Jiang, H., & Xu, H. (2018). Magnetic particles modification of coconut shell-derived activated carbon and biochar for effective removal of phenol from water. *Chemosphere, 211,* 962–969.
38. Thengane, S. K., Bandyopadhyay, S. (2020). Biochar mines: Panacea to climate change and energy crisis?. *Clean Techn Environ Policy, 22,* 5–10.
39. Rodrigues, T., & Braghini, A., Jr. (2019). Technological prospecting in the production of charcoal: A patent study. *Renewable and Sustainable Energy Reviews, 111,* 170–183.
40. Oliveira, A. C. (2012). *Sistema forno-fornalha para produção de carvão vegetal.* Universidade Federal de Viçosa.
41. García-Quezada, J., Musule-Lagunes, R., Prieto-Ruíz, J. A., Vega-Nieva, D. J., & Carrillo-Parra, A. (2023). Evaluation of four types of kilns used to produce charcoal from several tree species in Mexico. *Energies, 16*(1), 333. https://doi.org/10.3390/en16010333
42. Food and Agriculture Organization of the United Nations. (1987). *Simple technologies for charcoal making* (FAO Forestry Paper 41). Food and Agriculture Organization of the United Nations.
43. Osse, L. (1974). *Leña, carbón y carbonización.* IFONA-UNDP. FAO. ARG 70/536, Documento de Trabajo No. 15. (In Spanish).
44. Penedo, W. R. (1980). *Uso da madeira pare fins energéticos.* Fundação Centro Technológico de Minas Gerais. CETEC Publicações Técnicas, SPT-001. (In Portuguese).
45. Vahram, M. (1978). *Quality of charcoal made in the pit-tumulus.* University of Guyana/National Science Research Council, Charcoal Unit Laboratory Report.
46. Amer, M., & Elwardany, A. (2020). *Biomass carbonization.* IntechOpen. https://doi.org/10.5772/intechopen.90480
47. Amer, M., Nour, M., Ahmed, M., Ookawara, S., Nada, S., & Elwardany, A. (2019). The effect of microwave drying pretreatment on dry torrefaction of agricultural biomasses. *Bioresource Technology, 286,* 121400.
48. UNE-EN 18125. (2018). *Biocombustibles sólidos determinación del poder calorífico.* INTECO.
49. Ozyuguran, A., Akturk, A., & Yaman, S. (2018). Optimal use of condensed parameters of ultimate analysis to predict the calorific value of biomass. *Fuel, 214,* 640–646.

9

Wood in the Construction Industry: Combining Strength and Sustainability

9.1 Introduction

Wood is a material that can be used in various industrial, transportation, and construction products. Until now, wood remains an important material for continuously developing research and technology. Wood has many advantages, such as distinctive texture, attractive appearance, high strength-to-weight ratio, and good thermal insulation properties. Wood can also be combined with other materials as a composite with resin, concrete, and steel. Additionally, wood is the only construction material that does not contribute to greenhouse gas emissions and is completely renewable and recyclable. Wood, concrete and steel materials have different embodied energies and emission levels. The production of various construction materials has significant environmental impacts, primarily through CO_2 emissions. For instance, concrete production releases 159 kg of CO_2 into the atmosphere, while steel production contributes a substantial 1.24 tons of CO_2. Aluminium manufacturing has an even higher environmental footprint, emitting 9.3 tons of CO_2. In contrast, wood offers an environmentally beneficial alternative, as it absorbs a net 1.7 tons of CO_2 from the atmosphere, making it a more sustainable choice for reducing carbon emissions in construction [1]. As a natural material, wood is a complex material with varied properties and is sensitive to environmental conditions and physical and chemical loads. Wood is an anisotropic material with high strength in the direction of the grain and low strength perpendicular to the grain, which must be taken into account in the design of wood structures [2, 3].

Wood comes from various tree species that can be grouped into several categories, aiding in determining the most suitable wood for specific purposes and ensuring efficiency and durability in use. These categories include the country of origin, strength/mechanical properties, physical characteristics, application, and species and aesthetics. The category of origin refers to the area or habitat where the wood grows, such as tropical or temperate forests, with notable examples from Indonesia and Malaysia. Strength and

DOI: 10.1201/9781003617594-9

mechanical properties are critical for structural applications, with international standards ensuring safe and efficient use. Physical characteristics, including density, grain pattern, and texture, influence workability and appearance. The application category covers wood's intended uses, such as in construction, furniture making, or handicrafts. Species and aesthetics involve unique traits of wood types like teak, meranti, and pine, making them suitable for specific uses. Understanding these groupings allows for better decision-making in using wood for various purposes, maximizing its properties for efficiency and durability. Country of origin is also often used to determine the characteristics of certain woods, such as American oak, *maple, and teak. Mahogany* wood, *merbau wood, ironwood, agarwood, teak wood, red meranti wood, bangkirai wood,* and *gelam wood from Indonesia* [4]. In Malaysia, there are types of wood such as: *balau, kempas, kelat, resak, lime, keruing, mengkulang, light red meranti,* and *geronggang* [5]. In terms of strength, wood is classified based on its mechanical properties, including flexural strength, compressive strength, and tensile strength. International standards, such as those implemented in Malaysia, cover these various aspects to ensure the safe and efficient use of wood in construction. Applications of wood refer to its use in various fields, such as building construction, furniture making, and handicrafts. Wood species is also an important factor in classification, with different types of wood such as teak, meranti, and pine having unique characteristics that make them suitable for certain uses. Softwoods (such as pine) and hardwoods (such as oak) are used accordingly, with softwoods more commonly used in Europe due to their availability and lower cost.

9.2 Composition of Wood

The chemical content in wood depends on various factors, such as the part of the tree (root, trunk, or branch), type of wood (for example, normal, tension, or compression), geographic location, climate, and soil conditions. There are two main chemical components in wood, namely lignin (18–35%) and carbohydrates (65–75%) which are in the form of complex polymers, as well as a small portion of another mixture of organic and inorganic minerals (ash) around 4–10%. In general, wood contains about 50% carbon, 6% hydrogen, 44% oxygen, and small amounts of metal ions. In terms of carbohydrate components, wood consists of cellulose which ranges from 40 to 50% of the weight of dry wood, and hemicellulose which ranges from 25 to 35% [6].

9.2.1 Wood Components

Tree trunks have several important parts, such as bark, cambium, sapwood, and heartwood, each serving a specific function [7]. The cross-section of the

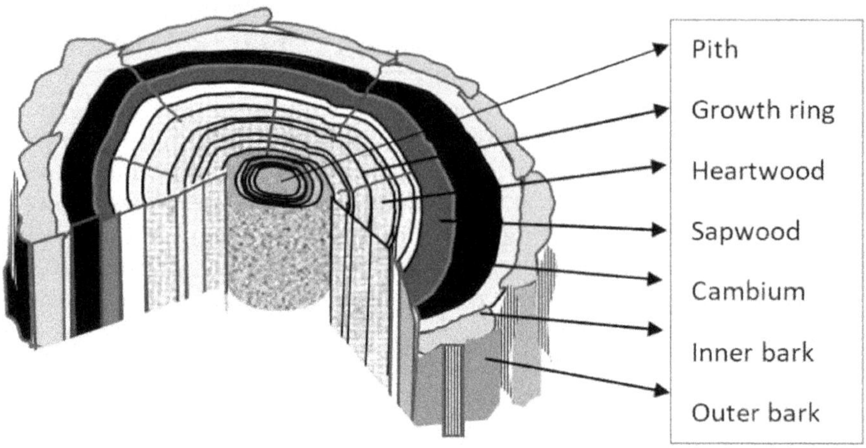

FIGURE 9.1
The cross-section of wood components [7].

stem in wood is characterized by distinctive annual growth rings (**Figure 9.1**), formed due to varying densities of wood at different times in the growing season. Most wood cells are tubular and elongated, known as fibres or tracheids. The cell walls have a multi-layered structure that provides strength and rigidity to the wood, consisting of various chemical components such as cellulose, hemicellulose, and lignin. Cellulose, the primary component, forms long fibres that provide high tensile strength. Hemicellulose and lignin fill the spaces between cellulose fibres and act as natural adhesives, holding these fibres together and providing stiffness and compressive strength to the wood. The direction of the cellulose fibres in these layers varies, enhancing the overall stability and strength of the wood. This unique structure of wood cell walls allows wood to possess a combination of strength, flexibility, and durability, making it a superior and versatile building material.

1. *Pith* (*heart of wood*): is the softest part of the wood but is very small in size compared to the diameter of the wood. This part should always be avoided and discarded.
2. Heartwood: the heartwood is the central, innermost part of the tree trunk. It is hard, dark in colour, and heavier than the outer layers. The heartwood forms as the tree ages, and older layers of sapwood gradually become heartwood. This transformation involves the deposition of various substances, such as tannins and other phenolic compounds, which provide natural resistance to decay and pests. As a result, heartwood is highly valued for its durability and strength, making it an essential component in many wood products.
3. *Sapwood*: the sapwood is the outer layer of the tree trunk, located just beneath the bark. It is lighter in colour compared to the heartwood

and plays a crucial role in the tree's life by transporting water and nutrients from the roots to the leaves. Sapwood is more prone to shrinking and swelling due to its higher moisture content.

4. *Cambium layer: the layer that contains food substances for tree development.*
5. *Bast: sends food to be processed by the leaves through photosynthesis.*
6. *Bark*: protects the tree trunk.
7. *Annular ring*: circular lines on a tree that indicate the age of the tree. The rings are formed each year based on the season in which the tree grows.
8. *Spring growth:* a layer that forms in spring. It is usually thinner because tree growth is slower during this season.
9. *Autumn growth:* a layer that forms in spring. It has more thickness because the trees grow faster during this season due to the processing of food for more trees.
10. *Medullary rays:* lines that run from the centre of the wood to the outside as a food storage medium for the tree. This part can be used as decoration when we cut round wood radially.

9.3 Physical Properties of Wood

- Density

Wood density in temperate climates varies between 0.3 to 0.9 g/cm^3. However, on a global scale, wood density generally ranges from 0.2 to 1.2 grams per cc. Variations in density among different species are mainly due to differences in wood composition, volume of voids within the wood, and the presence of extractive substances. The density of the wood itself, which does not include voids or extractive substances, is approximately 1.5 g/cm^3. These values are very consistent across different wood species, indicating that the basic structural components of wood are fairly uniform. The overall density of a particular piece of wood is influenced by several factors: wood components, cavity volume. The presence of pores, vessels, and other cellular structures can significantly influence overall density. Wood with larger or more voids will have a lower density [8].

- Thermal properties

Wood has low thermal conductivity and a low coefficient of thermal expansion (CTE). The degree of anisotropy in thermal properties is lower

TABLE 9.1
Grouping of Wood Durability

	Class 1: Very Durable	Class 2: Durable	Grade 3:	4th Grade:
In a protected room	>50 years	>50 years	50 years	50 years
Outdoor above ground	>50 years	30+ years	15 years	5-8 years
In the ground	>25 years	15–25 years	8–15 years	<5 years

compared to structural properties. The thermal conductivity parallel to the fibre is about two to three times greater than that perpendicular to the fibre. The average value for softwood is 1.2 W/mK. The CTE parallel to the fibre is about 10 to 20% of the value perpendicular to the fibre and usually ranges from 3–5 x 10^{-6} /°C [2] .

- Durability

Fungi, which use wood as a food source, cause rot which can result in a decrease in mechanical properties. This rot often affects the structural integrity of the wood, reducing its strength and causing significant damage if left untreated. Apart from fungi, insects such as termites and wood beetles can also damage wood by chewing and digging tunnels in the wood fibres. The presence of these insects can accelerate wood damage, especially if the wood is not given adequate protection. To increase the durability of wood, chemical processing and physical protection are often applied, such as the use of wood preservatives and protective coatings. These methods help extend the life of the wood and maintain its structural performance in the long term. This fungus requires moist conditions for growth and can be prevented by ensuring the moisture content of the wood is kept below 20% [2]. Wood durability class relates to the wood's natural ability to resist rot and insect attack. For wooden decks, pergolas, and handrails, it is recommended to use class 1 or 2 hardwood. In **Table 9.1,** various durability properties of wood are shown [9].

9.4 Mechanical Properties of Wood

- Strength

Understanding the strength of wood under various loads is crucial for its application in specific purposes, such as railway sleepers, house frames, bridges, and harbour structures. The bending, compression, and tensile properties of wood are governed by various international standards. The permissible design codes for stress vary from country to country; for instance, Malaysia follows MS 544, while the UK uses CP 112 or EN 1995-1-1, known as Eurocode 5, which has been adopted in the UK and most of Western Europe for at least the last five years. This paper aims to compare MS 544:2001 and Eurocode 5 in terms of design philosophy and methodology, highlighting the similarities and differences between the two codes of practice, particularly regarding flexible timber design [10, 11].

In Malaysia, the stress values for various wood strength classes are divided into 7 and coded SG1–SG 7, with SG1 being the class with the highest mechanical properties, often used for structural applications with high demands, for example, *Balau wood*. Wood with the SG1 code has bending stress: 14.0 MPa, shear stress: 1.4 MPa, stress parallel to the grain: 17.0 MPa, stress perpendicular to the grain: 2.5 MPa, and tensile stress parallel to the grain: 10.0 MPa [11]. Meanwhile, SG2–SG 6 is the middle class, suitable for general structural purposes, for example, *kempas, resak, kerung wood*. SG7 is a lower grade, usually used where high strength is not required, for example, *geronggong wood* [12]. In Indonesia, the strength properties of wood are divided into five strength classes, namely strength class I (very strong), strength class II (strong), strength class III (less strong), strength class IV (not strong), and strength class V (very not strong) [13]. Wood strength is affected by increasing water content; wood strength decreases linearly with increasing water content until it reaches 27% water content, which is the fibre saturation point (**Figure 9.2**). In particular, mechanical properties decrease by approximately 2–3% for every 1% increase in water content [14]. Further increases in moisture content have no effect on the strength or stiffness of

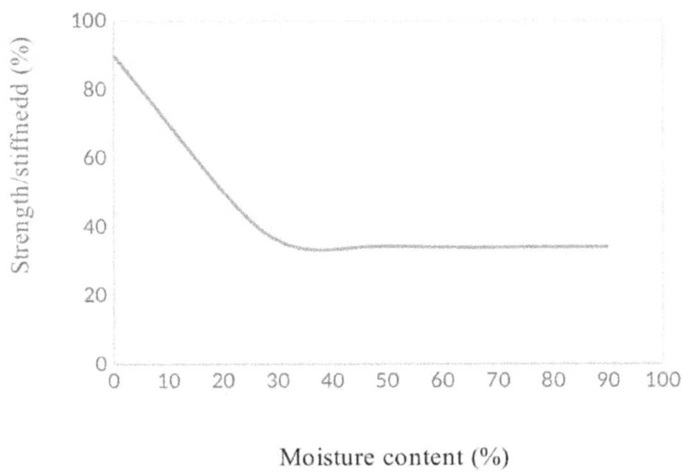

FIGURE 9.2
The relationship between decreasing strength/stiffness and the amount of water in wood [14].

TABLE 9.2

Effect of Grain Irregularities on the Strength Properties of Wood [15]

Slopes	Bending Strength (%)	Parallel Compressive Strength (%)	Impact Load (%)
Straight fibre	100	100	100
1 in 20 (3°)	93	100	95
1 in 10 (6°)	81	99	62
1 in 5 (11.5°)	55	93	36

the wood. It should be noted that although the pattern of change in strength and stiffness characteristics with changes in water content is similar for most mechanical properties, the magnitude of change differs from one property to another.

From a structural perspective, the direction parallel to the grain of wood is the strongest and stiffest. The tensile strength of softwood, when measured parallel to the grain at a moisture content of 12%, generally ranges from 70 to 140 MPa, while the compressive strength is typically lower, ranging from 30 to 60 MPa [2]. For hardwoods, these values are generally higher, indicating that wood can withstand significant loads when force is applied parallel to the grain direction. This property is particularly important in construction applications, where wood is used as beams, posts, and other structural elements requiring high tensile strength. Conversely, the properties of wood perpendicular to the grain are significantly lower than those parallel to the grain. The tensile strength in the radial and tangential directions can be as low as 5 to 8% and 3 to 5% of the tensile strength measured parallel to the grain, respectively. This indicates that wood is more susceptible to damage and deformation when loads are applied in these directions. **Table 9.2** shows the effect of fibre direction deviation on the strength properties of wood. This table compares compressive strength parallel to the grain, flexural strength, and impact load at various fibre inclinations.

9.5 Wood Defects

Wood has hygroscopic properties, which means that it can absorb water as it tries to achieve a balance of water content with the surrounding environment. Softwood is more susceptible to shrinkage than hardwood. The degree of shrinkage/swelling varies with direction. Values parallel to the grain direction are 5 to 10% of the values perpendicular to the grain direction, while values in the tangential direction can be one to two times that of the radial direction.

Because logs vary in cross-section along their length and usually taper towards one end, a board that is rectangular at one end may not be rectangular at the other end. This rectangular cross-section can intersect with the tapered exterior of the log, resulting in rounded edges. The impact of wood shrinkage is a reduction in cross-sectional area, which results in a decrease in strength properties and the formation of defects. These defects include [14]:

- Shake: cracks along the wood grain, often caused by uneven drying or internal stress.
- Knots: areas where tree branches once attached, creating circular or oval-shaped imperfections that can weaken the wood.
- Wane: the presence of bark or absence of wood on the edge or corner of a piece of wood, usually due to the shape of the log.
- End splitting: cracks that develop at the ends of pieces of wood, often as a result of rapid drying or mechanical damage.
- Honeycombing: internal cracks in wood that are not visible on the surface, usually caused by severe drying stress.
- Cupping: warping across the width of the wood, resulting in a curved or concave shape, usually due to uneven moisture content. Examples of conversion defects or natural defects in wood are shown in **Figure 9.3**.

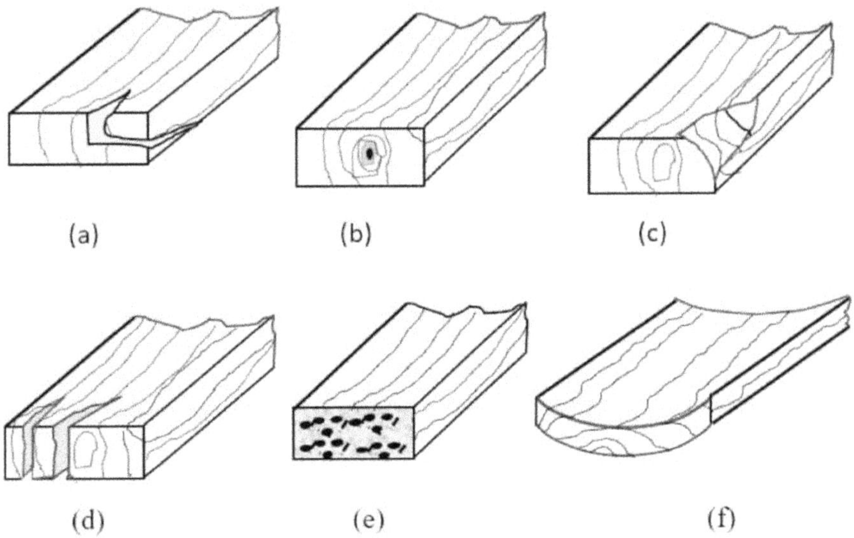

FIGURE 9.3
Types of defects in wood [14]. a: shake. b: knots. c: wane. d: end splitting e: honeycombing. f: cupping.

9.6 Engineered Wood

Engineered wood is a composite material crafted from processed or manufactured wood components. Examples of engineered wood products include plywood, particleboard, and laminated veneer lumber, as illustrated in **Figure 9.4**. This material is flexible and sustainable which has gained significant popularity in various engineering applications. *Engineered wood* is produced by combining natural wood fibres or particles with adhesives and other additives to increase its strength, durability, and dimensional stability. This process produces a versatile material that can be used in a variety of structural and non-structural applications, providing an environmentally friendly alternative [16, 17]. Types of *engineered wood* can be classified as follows.

- *Laminated board (plywood)*

Plywood is made by attaching thin layers of wood on top of each other. The mechanical properties of plywood can be varied by laying the layers at different angles using glue. Typically, a 45° angle is used between each layer to effectively withstand bending and twisting forces [18].

FIGURE 9.4
Various types of engineered wood, along with their distinct structures (Illustration: canva, 2024, https://www.canva.com).

- *Chipboard (particle board)*

Particle board or *chipboard* is made from wood waste such as sawdust and wood chips. This wood waste is converted into wood chips through a special process, then these wood chips are mixed with resin or glue during a thermal compression process. The final product of this process is called particle board or chipboard.

- MDF (*medium density fibreboard*)

MDF is made by combining sawdust, wood chips, or even organic fibres and then pressing them under high pressure. MDF, as a multi-layered board, is widely used as a building material in residential and commercial projects. The downside to MDF is that it is very high density, making it much heavier than plywood and particle board.

- HDF (*high-density fibreboard*)

High-density fibreboard (HDF) is a composite sheet made from pressed wood particles. HDF is a very thin sheet whose thickness generally varies between 3 and 8 mm. HDF is an artificial wood product produced from a combination of wood chips impregnated with synthetic resin and adhesive. The mixture of wood particles and resin is processed at high temperature and pressure with a thickness of less than 1 cm. The wood chips used in HDF are much more homogeneous than MDF and chipboard; therefore, their density is higher, about 900 kg/m^3 [19].

9.7 Wood Classification

Wood is classified into groups based on its strength and structural properties to ensure that each group has certain characteristics to guide its use. The methods used for this assessment include [20]:

- *Visual grading*: visual grading is a method for determining the stress level of wood, which is widely used in Australia and internationally, including in Europe and North America. In this process, a trained appraiser visually inspects each piece of wood against established standards for hardwood or softwood. These standards define the types, sizes, and positions of physical characteristics permitted within each structural level. The assessor compares the size and position of the nodes and other strength-reducing features to the allowable limits for the various assessment classifications. Higher levels allow fewer and smaller defects.

- *Machine stress assessment*: machine stress assessment uses machines to identify the strength of wood, measure the stiffness of the wood, and use the correlation between stiffness and strength to classify the stress level. Grouping woods based on narrow E (stiffness) ranges can result in a wider range of strengths. The E value also provides information about other structural properties, such as tensile, compression, and shear strength.
- *Stress grades:* wood stress level classification refers to grouping wood based on its mechanical properties, especially strength and stiffness characteristics. These stress levels are important in determining the suitability of wood for various structural applications, ensuring that it can safely support loads and withstand forces without failure.

The following are examples used to classify wood according to Australian Standards [21]:

- *AS/NZS 1748 Timber Stress-graded Product requirements for mechanically stress-graded timber.*
- *AS 2082 Visually stress-graded hardwood for structural purposes.*
- *AS 2858 Timber softwood - visually stress-graded for structural purposes.*
- *AS 2878 Timbers classification into strength groups.*
- *AS 3519 Timber Machine Proof Grading.*

In the UK, the required specifications are listed in BS 4978 and BS 5756 to determine whether a piece of wood is accepted in one of two visual stress levels or rejected. The following are the wood classification codes used in England:

- *BS EN 14081-1:2005. Timber Structures – Strength Graded Structural Timber with Rectangular Cross Section. Part 1: General Requirements, British Standards Institutions.*
- *7 BS EN 14081-2:2005. Timber Structures – Strength Graded Structural Timber with Rectangular Cross Section. Part 2: Machine Grading, Additional Requirements for Initial Type Testing, British Standards Institution.*
- *8 BS 4978:1996. Specification for Visual Strength Grading of Softwood, British Stan dards Institution.*
- *9 BS 5756:1997. Specification for Visual Strength Grading of Hardwood , British Standards Institution.*
- *10 BS 5268-2:2002. Structural Use of Timber. Part 2: Code of Practice for Permissibility Stress Design, Materials and Workmanship , British Standards Institution.*

In general, classifying different types of wood based on relative strength involves dividing them into strength groups. Since most types of wood become stronger when dried, there are two sets of strength groups: one for fresh wood, ranging from S1 (the strongest) to S7, and another for dried state; it is given the symbol D starting from D1 to D8. D1 refers to the density level of dry wood that has been processed. D1 is the level with the lowest density of dried wood. Wood at this level usually has a low density and is lighter. D8 is wood at the highest density level. Wood species that are typically used in special applications where extreme strength and density are required include hardwoods such as oak, hickory, and maple, which are often selected for high-stress applications like tool handles, flooring, and furniture.

Example National Standards in various countries including is as following [20]:

- *American Lumber Standards Committee. Standard Grading Rules for Northeastern Lumber; published by the Northeast Lumber Manufacturers Association (NeLMA).*
- *French Standard NF B 52-001. Regles d'utilisation du bois dans les constructions— Classement visuel pour l'emploi en structure pour les principales essences résineuses et feuillues.*
- *German Standard DIN 4074 Teil 1. Sortierung von nadelholz nach der tragfahigkeit, nadelschnittholz.*
- *Irish Standard IS I27. Specifications for stress grading softwood timber.*
- *Italian Standard UNI 11035-1/-2. Structural timber—Visual strength grading for structural timbers.*
- *Japanese Standard JAS 143. Structural softwood lumber.*
- *Japanese Standard JAS 600. Structural lumber for wood frame construction.*
- *Korean Standard KS F 2151. Visual grading for softwood structural lumber.*
- *Netherlands Standard NEN 5493. Quality requirements for hardwoods in civil engineering works and other structural applications.*
- *Netherlands Standard NEN 5499. Requirements for visually graded softwood for structural applications.*
- *New Zealand NZ S 3631. Timber Grading Rules.*
- *Slovenian Standard SIST DIN 4074-1. Strength grading of wood—Part 1: Coniferous sawed timber.*
- *Spanish Standard 56544. Visual strength grading for structural sawn timber. Softwood timber.*
- *South African Standard SABS 1783. Sawn Softwood Timber.*

9.8 Wood Treatment

Wood treatment involves the chemical or physical modification of wood to enhance its durability and resistance to decay, ensuring it meets desired performance characteristics. Chemical treatments are commonly used to improve wood's resistance to fire and rot during its service life. Each type of treatment and chemical has a specific effect on the mechanical properties of wood [22]. When determining whether wood treatment is necessary, it is important to consider several factors: the presence of hazardous materials, potential harm to the environment, the cost of failure, the estimated service life of the structure, and the natural resistance of the wood. Here are some approaches to wood treatments:

- High pressure treatment: this method involves pressing a chemical preservative into the wood fibres using high pressure. This allows the density of the wood to increase [23, 24].
- Vacuum treatment: the wood is placed in a vacuum chamber where the air is sucked out before a chemical preservative is injected. This vacuum process allows for more even and efficient curing.
- Heat treatment: wood is heated at high temperatures to reduce its water content and increase resistance to attack by decomposing organisms. Heat treatment of wood is usually carried out at temperatures between 160 and 240°C, resulting in wood that is darker in colour, with better dimensional stability and increased microbial resistance, although strength is reduced, especially in terms of resistance to fracture [25]. This treatment is considered successful if carried out above 150°C, where thermal degradation processes occur [26].
- Surface treatment: wood treatment is carried out to protect the surface of the wood from biotic and abiotic factors, including UV rays, rainwater, and moisture, through the application of a protective coating. These coatings enhance the wood's durability against weathering. Additionally, wood veneers can be used for aesthetic purposes, allowing the wood's natural beauty to remain visible while prolonging its service life. The working mechanism of wood coatings can be categorized into two types: film formation and penetration [27]. Penetrating coatings include water repellents, transparent stains, preservatives, and surface treatments.
- Oil-based preservatives: examples of oil-based preservatives are polycyclic aromatic hydrocarbons (PAH), creosote (polycyclic aromatic hydrocarbons and phenols), pentachlorophenol (PCP/penta), copper naphthenate. These preservatives have been used in various

American and European countries to coat wooden bridges, railway sleepers, and building structures [2–4].

- Water-based preservatives: water-based wood treatments offer several advantages, including cost-effectiveness, safety, and the ability to provide a clean surface on the treated wood [28]. Wood preserved with water-based chemicals can be painted after treatment and is suitable for a wider range of applications, such as utility poles, residential wood, and composite wood protection. These treatments are effective in preventing or limiting strength reduction associated with chemical and preservative-based processes. Examples of water-based wood treatments include copper, which, in higher doses, acts as a preservative against rot, bacteria, fungi, insects, and algae [29]. Boric preservatives, which are water-soluble, have also been shown to be effective against wood-destroying insects and fungi [30].
- Natural wood preservatives: natural wood preservatives can be categorized based on the origin of the compounds including plant extracts, essential oils, waxes, resins, tannins from bark, wood core, and chitosan. Antimicrobial substances are aromatic and non-aromatic compounds produced by plants, including phenols, terpenoids, alkaloids, lectins, and polypeptides that can be used in various applications. Several studies have investigated derivatives from various plant parts, including bark, wood, leaves, seeds, and fruit which can be used for wood protection. For example, extracts from cinnamon leaves have been shown to be effective against wood-rotting fungi and termites [31, 32]. Extracts from cinnamon leaves have also been shown to be effective against wood rot, fungi, and termites [33].
- Anti-fungus and insects: the following wood preservatives are commonly used to fight fungi and insects effectively: boron compounds, carbamates, copper inorganic compounds, copper naphthenates and citrates, creosote, isothiazolones, quaternary ammonium compounds, sulphamides, and triazoles.

9.9 Nondestructive Tests

Nondestructive test (NDT) is a method to determine the physical and mechanical properties of materials without damaging or changing their properties before testing. Thus, after testing, the material properties remain the same as before testing. There are various types of nondestructive testing (NDT), and their use depends on the specific purpose of the test. The primary aim of nondestructive testing is to provide accurate information regarding the properties, performance, or condition of a material when subjected to external loading or influences. Below are several examples of NDT methods:

Static Bending Methods: This method aims to determine the modulus of elasticity (MOE) of wood. MOE is also referred to as the modulus of elasticity, as the deflection of objects results from shear loads and bending moments [34].

Longitudinal Stress Wave Method (LSWM): This method is used to obtain values for the dynamic elastic modulus (Ed). It employs a tool that records impulses produced when wood is struck with a hammer, helping to identify the resonant frequency (f_0). The dynamic elastic modulus is calculated using the values of f_0, the length of the wood (L), and the wood density (ρ) [35].

X-ray CT Scanning: This method, which uses ionizing radiation (X-rays or gamma rays), relies on the principle that each material has different photon absorption energies. By measuring the reduction or attenuation of the energy source as it passes through the specimen, density variation maps of the inhomogeneous internal structure can be created [36].

Infrared Thermal Imaging Method: This testing method uses a thermal camera to produce a thermal map of the material, which is then compared to a colour reference. This approach allows for the detection of thermal issues occurring in the inspected area [37].

References

1. Kalimantan Recycled Timber. (2017). https://www.kaltimber.com/blog/2017/6/19/how-much-co2-is-stored-in-1-kg-of-wood
2. Harte, A. (2009). Introduction to timber as an engineering material. In M. Forde (Ed.), *ICE manual of construction materials*, 707–715. National University of Ireland Galway.
3. Ahammed, Md R., Mia, M. B., Raihan, Md A., Hossain, Md N., Hossen, M., Hasan, M., & Shahriar, M. (2024). An overview of conventional construction materials and their characteristics creators. *North American Academic Research*, 7(1),1–13. https://doi.org/10.5281/zenodo.10579091
4. Basri, E., Azmi, A., Ahmad, Z., Lum, W. C., Baharin, A., Za'ba, N. I. L., Bhkari, N. M., & Lee, S. H. (2022). Compressive strength characteristic values of nine structural sized Malaysian tropical hardwoods. *Forests*, *13*(8), 1172. https://doi.org/10.3390/f13081172
5. European Committee for Standardization. (2016). *EN 338:2016; Structural timber-strength classes*. European Committee for Standardization.
6. Pettersen, R. C. (1984). The chemistry of solid wood. In *Advances in Chemistry series* (Vol. 20, pp. 57–126). ACS.

7. KidsPressMagazine. (n.d.). *Wood composition*. https://kidspressmagazine.com/science-for-kids/misc/misc/wood-composition.html
8. Britannica. (n.d.). *Properties of wood*. Retrieved from https://www.britannica.com/science/wood-plant-tissue/Properties-of-wood
9. European Committee for Standardization. (2016). *EN 384:2016; Determination of characteristic values of mechanical properties and density*. European Committee for Standardization.
10. Sheikh, S., & Ahmad, Y. (2015). Flexural timber design to Eurocode 5 and the Malaysian Timber Code Ms 544: 2001. *Malaysia Journal of Civil Engineering, 27*(1), 207-223.
11. Department of Standards Malaysia. (2017). *MS 544: Part 2: 2017; Code of timber practice for structural use of timber: Part 2: Permissible stress design of solid timber*. Department of Standards Malaysia.
12. Nur Aidi, M. (2011). Pengelompokan kelas kekuatan pada beberapa jenis kayu berdasarkan sifat mekanik dengan menggunakan analisis gerombol. *Physics Materials Science Composite Material, 6*(4), 25-36.
13. Md Noh, N. I. F. (n.d.). *Timber as a structural material 1.1*. https://www.academia.edu/23111939/Timber_As_A_Structural_Material_1_1_Introduction
14. Ozyhar, T., Hering, S., & Niemz, P. (2012). Moisture-dependent elastic and strength anisotropy of European beech wood in tension. *Journal of Materials Science, 47*, 6141–6150. https://doi.org/10.1007/s10853-012-6534-8
15. Zhang, J. (2023). Current applications of engineered wood. In *Intec*. https://doi.org/10.5772/intechopen.112545
16. Bayat, M., & Abootorabi, M. M. (2021). Comparison of minimum quantity lubrication and wet milling based on energy consumption modeling. *Proceedings of the Institution of Mechanical Engineers, Part E: Journal of Process Mechanical Engineering, 235*(5), 1665–1675.
17. Baharin, A., Fattah, N. A., Bakar, A. A., & Ariff, Z. M. (2016). Production of laminated natural fibre board from banana tree wastes. *Procedia Chemistry, 19*, 999–1006.
18. Young, T. M., et al. (2020). Improving innovation from science using kernel tree methods as a precursor to designed experimentation. *Applied Sciences, 10*(10), 3387.
19. Bayat, M. (2023). Types of engineered wood and their uses. In Jun Zhang (Ed.) *Current applications of engineered wood*. Intetech, 1-20.
20. Arriaga, F., Wang, X., Íñiguez-González, G., Llana, D. F., Esteban, M., & Niemz, P. (2023). Mechanical properties of wood: A review. *Forests, 14*(6), 1202. https://doi.org/10.3390/f14061202
21. Mohamad Omar, M. K., & Mohd Jamil, A. W. (2012). Use of timber in engineering. In S. Vimala (Ed.), *Timber technology bulletin, ministry of natural resources and environment, Malaysia* (p. 50). Forest Research Institute Malaysia 50, 1-11.
22. Jennings, J. D. (2003). *Investigation the surface energy and bond performance of compression densified wood* (Master's thesis). Virginia Polymeric Institute and State University, Blacksburg, VA.
23. Khademibami, L., & Bobadilha, G. S. (2022). Recent developments studies on wood protection research in academia: A review. *Frontiers in Forests and Global Change, 5*. https://doi.org/10.3389/ffgc.2022.793177
24. Bekhta, P., & Niemz, P. (2003). Effect of high temperature on the change in color, dimensional stability and mechanical properties of spruce wood. *Holzforschung, 57*, 539–546. https://doi.org/10.1515/HF.2003.080

25. Yin, Y., Berglund, L., & Salmen, L. (2011). Effect of steam treatment on the properties of wood cell walls. *Biomacromology, 12*, 194–202. https://doi.org/10.1021/bm101144m
26. Williams, R. S. (1999). Finishing of wood. In R. J. Ross (Ed.), *Wood handbook: Wood as an engineering material*. U.S. Department of Agriculture.
27. Smith, S. T. (2019). 2018 railroad tie survey. *Journal of Transportation Technology, 9*, 276. https://doi.org/10.4236/jtts.2019.93017
28. Preston, A., Jin, L., Nicholas, D., Zahora, A., Walcheski, P., Archer, K., et al. (2008). *Field stake tests with copper-based preservatives*. Paper presented at the 39th Annual Meeting of the International Research Group on Wood Protection (IRG), Istanbul.
29. Williams, L. H. (1996). Borate wood-protection compounds: A review of research and commercial use. *APT Bulletin: Journal of Preservation Technology, 27*, 46–51. https://doi.org/10.2307/1504564
30. Broda, M. (2020). Natural compounds for wood protection against fungi-a review. *Molecules, 25*, 3538. https://doi.org/10.3390/molecules25153538
31. Yang, D. Q. (2009). Potential utilization of plant and fungal extracts for wood protection. *Forest Products Journal, 59*, 97–103.
32. Cheng, S. S., Liu, J. Y., Hsui, Y. R., & Chang, S. T. (2006). Chemical polymerization and antifungal activity of essential oils from leaves of different provenances of indigenous cinnamon (Cinnamomum osmophoeum). *Bioresource Technology, 97*, 306–312.
33. WoodSolutions. (n.d.). *Timber preservation*. https://www.woodsolutions.com.au/timber-preservation
34. American Society for Testing and Materials. (2021). *ASTM D143:2021; Standard test methods for small clear specimens of timber*. American Society for Testing and Materials.
35. Utami, S. S., et al. (2020). Design of an acoustic-based nondestructive test (NDT) instrument to predict the modulus of elasticity of wood. *Engineering Journal, 24*(6), 109–116.
36. Lindgren, L. O. (1991). The accuracy of medical CAT-scan images for non-destructive density measurements in small volume elements within solid wood. *Wood Science and Technology, 25*, 425–432.
37. The Constructor. (n.d.). *8 advanced non-destructive testing methods*. https://theconstructor.org/exclusive/8-advanced-non-destructive-testing-methods/294295/#goog_rewarded

10

Future Opportunities in Green Engineering Materials

10.1 The Future Energy-Efficient Materials in Green Building Construction

The future of green engineering materials for construction (green building) will focus on the integration of sustainable materials in both design and construction processes [1, 2]. Green buildings prioritize energy efficiency, water conservation, improved air quality, and the use of renewable and recycled materials. With their focus on cost savings, energy optimization, and harmony with nature, green buildings are becoming a cornerstone of modern sustainable development. The principles of green building emphasize the integration of energy efficiency and sustainable material selection throughout the building's lifecycle. Key components of green materials include considerations for their manufacturing processes, building operations, and waste management strategies [3]. Sustainable practices aim to minimize pollution, treat and conserve water, and incorporate renewable energy sources. Green materials prioritize attributes like non-toxicity, reusability, reduced embodied energy, and the use of natural materials. They are also designed for longer lifespans, contributing to environmental conservation and resource efficiency while meeting modern building standards [4, 5].

Green building offers significant advantages across environmental, economic, and social aspects. Environmentally, it improves water and air quality, reduces wastewater, protects ecosystems, and conserves natural resources. Economically, green buildings lower operational costs, enhance occupant productivity, and create new markets for green services and products. Socially, they improve the quality of life, reduce strain on local infrastructure, and provide healthier, more comfortable living environments for occupants. By integrating these advantages, green building practices contribute to a more sustainable and resilient built environment [6–8].

In many developing countries, the construction sector significantly affects the environment. Buildings account for 40% of the natural resources used, 70% of the electricity consumed, and 12% of drinkable water. They also

DOI: 10.1201/9781003617594-10

produce 44–64% of the solid waste, much of which ends up in landfills. This has contributed to a high level of emissions, with buildings being responsible for 30% of greenhouse gases, including 18% from the extraction and transportation of materials. Additionally, building construction consumes 25% of the world's timber, causing a major impact on global forests. These statistics underscore the substantial ecological footprint of the construction industry [9].

In response to the environmental challenges presented by the construction industry, Europe's green building programme has set ambitious energy efficiency goals, requiring buildings to reduce primary energy demands by 25% to 50% compared to conventional standards [10]. Focusing primarily on non-residential buildings such as offices, schools, swimming pools, and industrial facilities, this initiative promotes energy-efficient practices and plays a crucial role in advancing sustainable development. Global rating systems also focus on environmental concerns by assessing sustainable buildings based on criteria such as energy efficiency, water conservation, material selection, and indoor environmental quality. These systems set benchmarks for the design, construction, and operation of buildings, ensuring certification of their sustainability performance. They assess key sustainability aspects like site development, human health, and environmental impact, offering quality assurance to both building owners and users [9]. The certification process aims to be user-friendly and reliable, facilitating clear communication of results and ensuring transparency throughout the assessment. The energy performance certificate (EPC), implemented by the EU, is a key certification for buildings, mandating energy-saving regulations since 2007. In Germany, the energy saving regulation sets maximum values for primary energy demand and heat loss in both residential and non-residential buildings, with the maximum values for renovations generally being 40% lower than those for new construction. The EPC aims to standardize energy performance across buildings, contributing to the EU's energy efficiency goals. Similarity, green building certifications like LEED, established by the US Green Building Council, have significantly driven global adoption of sustainable construction practices [10]. The LEED certification evaluates buildings based on sustainability metrics, encouraging environmentally conscious construction. Egypt introduced the green pyramid rating system to assess and guide eco-friendly building practices, demonstrating a growing international commitment to environmentally responsible building methods. Together, these efforts highlight a worldwide shift towards sustainability in the built environment [11].

One significant development in green building is the use of municipal solid waste incineration (MSWI) ash, which can be recycled for various applications in construction. MSWI residues are increasingly utilized to partially replace light aggregates or sand in building projects. This practice has gained momentum in the European construction industry, with countries such as France, Germany, Denmark, and the Netherlands recycling over 70% to 90% of MSWI ash in their construction efforts. As a result, research on improving the applicability, performance, and pollution reduction of MSWI has seen substantial growth, highlighting its potential as a sustainable building material [12–14].

In parallel, technology plays a pivotal role in advancing green construction practices. Building information modelling (BIM) enables precise planning that optimizes resource use and reduces waste, while smart building systems, powered by IoT devices, monitor and automate energy usage for enhanced efficiency. Furthermore, energy balancing in green buildings goes beyond reducing heat loss during transmission [15]. It also accounts for factors like solar radiation as a heat source, internal heat generation, and energy distribution, storage, and transfer within the building to improve overall energy performance.

Smart building technologies and the Internet of Things (IoT) transform building operations by enabling real-time monitoring and management of energy, water usage, and indoor air quality. Smart sensors and energy management systems optimize resource use, reducing operational costs and carbon footprints. Renewable energy technologies, such as solar panels and wind turbines, can be seamlessly integrated into building designs to support on-site energy generation [16, 17]. Moreover, material advancements, like self-healing concrete, bio-based insulation, and recycled composites, are facilitated by breakthroughs in material science and nanotechnology. These innovations extend the lifespan of structures, further reducing the environmental impact of building materials over time. By bridging the gap between sustainability and functionality, technology not only makes green construction feasible but also drives its efficiency and appeal, paving the way for a more sustainable future in the built environment [18].

10.2 Eco-friendly Innovations: Advancing Green Product Development

Green technology encompasses a wide range of innovations designed to minimize environmental impact and promote sustainability across various sectors. Examples include renewable energy technologies such as *solar panels, wind turbines, and hydropower systems,* which provide clean and efficient energy. *Solar panels* are a renewable energy technology that converts sunlight into electricity through the *photovoltaic (PV) effect* [19–21]. *Wind turbines* are devices that convert the kinetic energy of wind into electrical energy through mechanical processes. They are a critical component of renewable energy systems and are widely used in wind farms to produce clean and sustainable power. They are a cornerstone of sustainable energy solutions due to their ability to harness clean and abundant solar energy.

Green roofs, sustainable materials, and *IoT*-enabled systems are pivotal innovations for advancing urban sustainability [22]. *Green roofs,* with their vegetation-covered surfaces, filter rainwater, reduce runoff, and provide thermal insulation while improving air quality and mitigating the urban heat island effect. Sustainable materials, such as laminated timber, offer eco-friendly alternatives to conventional concrete and steel, cutting down carbon emissions during production and providing robust structural performance. Meanwhile, IoT-enabled systems optimize energy usage in buildings by monitoring consumption in real time, automating lighting, HVAC systems, and other utilities based on occupancy. Together, these technologies and materials significantly reduce environmental impact while enhancing the functionality and efficiency of urban infrastructure [23].

Energy-efficient solutions like LED lighting and smart grids reduce consumption and waste [24]. Residential LEDs consume at least 75% less energy and last up to 25 times longer than incandescent lighting. Widespread use of LED lighting has a large potential impact on energy savings in the United States. By 2035, the majority of lighting installations are anticipated to use LED technology, and energy savings from LED lighting could top 569 TWh annually by 2035, equal to the annual energy output of more than 92 1,000 MW power plants [25]. Modern *LED* systems can integrate with smart home technologies, enabling automated control, dimming, and colour adjustments, further enhancing energy conservation and user convenience. *Low-emissivity (Low-E) glass* plays a vital role in enhancing resource conservation by improving energy efficiency in buildings [26]. *Low-E glass* minimizes heat transfer, helping to maintain indoor temperatures. This reduces the need for heating in winter and cooling in summer, significantly lowering energy usage. Waste management technologies, including recycling plants and composting systems, focus on reducing landfill usage and repurposing waste. Water conservation is addressed through *greywater recycling* and *rainwater harvesting systems,* while advanced manufacturing methods like *3D printing* and *biodegradable material* production minimize resource usage [27]. Transportation innovations such as *electric vehicles* and *hydrogen fuel cells* further reduce greenhouse gas emissions. These technologies collectively support a sustainable future by optimizing resource use and reducing environmental degradation [28].

In bioengineering, the use of sustainable materials with high performance is also a critical focus. These materials, derived from biological sources or designed to integrate with natural systems, offer key properties such as strength, thermal stability, and durability [29, 30]. Examples include bamboo and hemp in construction, cork for insulation, and soy-based foams for eco-friendly consumer products [31]. Biocomposites, such as flax- or jute-based resins, and alternatives like PLA (polylactic acid), are driving advancements in industries like automotive, packaging, and 3D printing, showcasing their role in energy-efficient and low-waste production processes [32]. The advancement of bioengineering is exemplified by the invention of biomaterials, enabling the restoration and enhancement of organ and tissue functions. This progress is driven by the development of innovative biomaterials and advancements in biofabrication techniques. These materials are meticulously designed to be compatible with specific medical application sites, ensuring non-toxicity while facilitating safe integration with biological systems. In tissue engineering, polymers derived from *lactic acid, glycolic acid,* and their copolymers are highly regarded for their outstanding biocompatibility and bioresorption characteristics [33, 34]. These properties allow the materials to naturally degrade within the body over time, minimizing long-term complications. Such innovations are instrumental in supporting the regeneration of damaged tissues and unlocking new possibilities for advanced medical treatments and improved patient outcomes.

10.3 Advanced Technologies in Green Development

10.3.1 Nanofabrication

Nanofabrication, which involves designing materials with precise properties at the nanoscale, plays a crucial role in advancing engineering materials with enhanced strength, flexibility, and functionality [35]. This technology is instrumental in creating high-performance materials for various applications. In engineering, nanofabrication is enabling the development of stronger, lighter, and more versatile materials. In fields like medicine, it is revolutionizing applications such as drug delivery systems, tissue engineering scaffolds, and biocompatible sensors [36, 37]. By manipulating materials at the molecular level, researchers gain greater control over material properties, paving the way for the development of sustainable, efficient, and high-performance bioengineering materials. These advancements are contributing to significant breakthroughs in medical treatments, diagnostics, and beyond.

The fabrication of micro and nano structures in the semiconductor industry significantly supports green industrial practices. Processes such as thin-film deposition and lithographic patterning enable the creation of energy-efficient devices like solar cells, LED lighting, and advanced sensors. Innovations in miniaturization and lithographic techniques ensure precisely defined geometries, reducing material usage and energy consumption during production. For instance, nanostructured photovoltaic cells enhance solar energy conversion efficiency by maximizing light absorption. These advancements not only minimize waste and optimize resource utilization but also offer time-efficient and cost-effective solutions, contributing to a reduced environmental footprint [38–40].

10.3.2 Smart Materials

Smart materials are a category of advanced materials that exhibit responsive behaviour to external environmental factors like temperature, pressure, light, or electrical fields [41]. These materials are designed to change their properties in real time, making them highly valuable in reducing energy consumption and improving the sustainability of various products, especially in construction and product design [42].

Shape-memory materials (SMAs) have gained attention for their potential in various green applications due to their ability to return to their original shape after deformation in response to external stimuli, such as heat. This property can be utilized in a range of eco-friendly applications, from self-healing materials to energy-efficient systems [42, 43]. Thermochromic materials, which change colour based on temperature, offer another promising green technology. These materials can act as real-time indicators of temperature fluctuations, eliminating the need for additional sensors or energy inputs. Such materials can be used in building materials, clothing, and packaging, contributing to energy conservation by signalling when temperature thresholds are reached, thus reducing the need for unnecessary energy consumption [44, 45]. Together, SMAs and thermochromic materials represent innovative approaches that combine functionality with sustainability, offering solutions that help reduce energy use, minimize waste, and promote eco-friendly design.

Another example of smart materials with significant potential for energy efficiency is self-healing materials [46]. These materials are designed to repair themselves after damage, such as cracks in structural components, by using embedded microcapsules or vascular networks filled with healing agents. In construction, self-healing concrete can help extend the lifespan of infrastructure by reducing maintenance needs and improving structural integrity. The ability to self-repair prevents energy losses associated with degradation, offering long-term cost benefits and contributing to the durability of buildings and infrastructure projects [47, 48].

10.4 Green Technology in Manufacturing Companies

Large amounts of domestic and industrial waste are a growing concern globally, threatening ecosystems and human health. Many products are discarded, rejected, or abandoned without proper management, highlighting the urgent need for green engineering solutions. Green technology has transformed manufacturing by integrating energy-efficient machinery, renewable energy sources, and waste recycling systems to enhance sustainability and performance. Companies adopt innovations like lean production to minimize material and energy waste, zero-waste manufacturing to recycle materials, and biodegradable packaging to reduce environmental impact [49]. Renewable energy solutions, such as solar panels and wind turbines, further lower dependency on fossil fuels while reducing operational costs and aligning with global sustainability goals.

Industry 4.0 technologies, such as visual computing, cyber-physical systems (CPS), and the Internet of Things (IoT), are transforming smart manufacturing by enhancing productivity, safety, and environmental sustainability. These technologies enable human-machine interfaces to optimize processes and address challenges like inventory management [50]. Big data analytics (BDA) is central to this transformation, enabling real-time data collection, storage, and analysis for improved decision-making, risk management, supply chain optimization, and environmental compliance [51]. Additionally, BDA supports logistics operations and remanufacturing, fostering operational efficiency and sustainable practices [52].

By integrating these advanced technologies, smart green manufacturing promotes strategies that reduce waste, enhance performance, and align with sustainability goals. Collectively, these innovations redefine manufacturing systems, driving efficiency, compliance, and a competitive edge while supporting a sustainable future.

10.5 Green Engineering with Industry 4.0

The Fourth Industrial Revolution (4IR) is recognized as a pivotal tool for advancing industrial capabilities. It enhances the flexibility, agility, and speed of production systems through innovative, efficient, and cost-saving technologies. The Fourth Industrial Revolution (4IR) integrates cyber-physical systems (CPS) with technologies such as IoT, big data analytics, robotics, additive manufacturing, machine learning, AI, blockchain, and automation. This advancement promotes cost-effective, labor-friendly, and environmentally sustainable operations. Furthermore, these technologies contribute to reduced energy consumption, lower emissions of toxic gases, and substantial waste reduction [53–55]. These technologies play a pivotal role in transforming production processes to meet real-time demands effectively [56]. Industry 4.0 technologies influence business models, organizational strategies, and production systems [57, 58].

Cyber-physical systems streamline operations such as scheduling and job execution, which reduces resource consumption and operational costs [59].

Moreover, data science, machine learning, big data analyses and advanced data analytics empower researchers to analyse extensive datasets, gaining insights into material properties, predicting material behaviour, and optimizing manufacturing processes. These methods enable lifecycle analysis for green materials like bioplastics and sustainable composites, assessing their environmental impact comprehensively [60,61]. Additionally, data science supports the integration of renewable resources and energy-efficient designs by utilizing data from climate models, material databases, and environmental impact reports. This fosters innovation, reduces waste, and strengthens circular economy practices. Leveraging these techniques ensures that green materials become more efficient and impactful in addressing global sustainability challenges [60,62].

In the service sector, the adoption of eco-friendly practices is exemplified by green banking, including eco-friendly products, services, and processes implemented by financial institutions [63]. Artificial intelligence (AI) plays a transformative role in this sector, revolutionizing banking globally by emulating human behaviour and supporting everyday activities [64]. AI-driven innovations in banking include online bill payments and internet banking, all of which reduce paper use and energy consumption, minimizing the carbon footprint of banking operations. For instance, online account openings eliminate paper-intensive procedures, enhancing customer convenience while also promoting environmental sustainability [65]. The benefits of green banking extend beyond environmental protection. By cutting down on paper usage and energy consumption, banks can realize substantial cost savings. Solar-powered ATMs exemplify this innovation by providing reliable financial services while reducing reliance on non-renewable energy sources. These advancements contribute to resource efficiency and support a greener, more sustainable banking system [66].

10.6 Economic Implications of Green Engineering Transition

Growing awareness of climate change has spurred the development of the global green finance market, projected to expand from USD 3,192.61 billion in 2023 to USD 22,754 billion by 2033, with a compound annual growth rate (CAGR) of 21.7% during 2024–2033 [67]. This market includes funding initiatives with environmental impacts, such as investments in renewable energy, sustainable agriculture, and carbon reduction activities. Financial products like green bonds, loans, and equities are vital tools for financing environmentally friendly projects, further advancing the transition to a sustainable economy.

The transformation towards the adoption of green materials and technologies has influenced green finance and sustainable economic development, including capital markets. Research, such as that conducted by Zhang et al. [68], highlights the positive impact of renewable energy investments on China's green economy. On the other hand, it highlights the balance between environmental protection and economic growth, showing that green finance plays a crucial role in reducing emissions [69]. Additionally, further evidence analyzing data from 44 countries demonstrates the substantial and long-lasting impact of green finance on the development of renewable energy [70].

The successful implementation of green engineering is strongly influenced by policies, which are crucial for accelerating the growth of a green economy. For instance, monetary policies have been shown to positively impact low-carbon industries through mechanisms like credit creation, particularly benefiting emerging economies [71]. Additionally, green policies, such as environmental taxes, play a significant role in promoting renewable energy investments. Furthermore, the extent of green regulation directly affects the relationship between green finance and renewable energy projects [72]. Green finance, in conjunction with policies like environmental regulations, green subsidies, and carbon taxes, has demonstrated significant benefits in reducing carbonization [73]. To support green economic activities, several countries and corporations have taken concrete steps. For example, the US Department of Energy's Loan Programs Office has committed over USD 40 billion in loans for renewable energy projects, carbon capture, and hydrogen, aiming to accelerate the transition to a clean energy economy. Similarly, in Europe, the EU's Invest EU programme is a major catalyst for green finance, forecasting at least €1 trillion in investments, with €260 billion per year required by 2030. Meanwhile, the corporate sector is also playing a vital role in the growth of green finance, as demonstrated by a record USD 500 billion in green bond issuances in 2021 [67].

References

1. Sharma, S., & Sharma, N. K. (2022). Advanced materials contribution towards sustainable development and its construction for green buildings. *Materials Today Proceedings, 68,* 968–973. https://doi.org/10.1016/j.matpr.2022.02.103
2. Sangmesh, B., Patil, N., Jaiswal, K. K., Gowrishankar, T. P., Selvakumar, K. K., Jyothi, M. S., Jyothilakshmi, R., & Kumar, S. (2023). Development of sustainable alternative materials for the construction of green buildings using agricultural residues: A review. *Construction and Building Materials, 368,* 130457. https://doi.org/10.1016/j.conbuildmat.2023.130457
3. Udechukwu, C. E., & Johnson, O. O. (2018). The impact of green building on valuation approaches. *Lagos Journal of Environmental Studies, 6*(1), 3–13.
4. Green roofs: A critical review on the role of components, benefits, limitations, and trends. (2019). *Renewable and Sustainable Energy Reviews, 57,* 740–755. https://doi.org/10.1016/j.rser.2015.12.108

5. Wahan, Y., Yanch, Y., & Jun, H. (2019). Elementary introduction to the green management of the construction in the whole process. *Physics Procedia, 24*, 1081–1085.
6. Katinas, V., Markevicius, A., Perednis, E., & Savickas, J. (2014). Sustainable energy development - Lithuania's way to energy supply security and energetics independence. *Renewable and Sustainable Energy Reviews, 30*, 420–428. https://doi.org/10.1016/j.rser.2013.10.004
7. Korol, S., Shushunova, N., & Shushunova, T. (2018). Innovation technologies in green roof systems. *MATEC Web of Conferences, 193*, 1–8. https://doi.org/10.1051/matecconf/201819300001
8. Yu, H. (2021). Environment sustainable acoustic in urban residential area. *Procedia Environmental Science, 10*, 471–477.
9. Abdelfattah, A. F. (2020). Sustainable development practices and its effect on green buildings. *IOP Conference Series: Earth and Environmental Science, 410*(1), 012025. https://doi.org/10.1088/1755-1315/410/1/012025
10. Bauer, M., Mösle, P., & Schwarz, M. (2009). *Green building – guidebook for sustainable architecture*. Springer Science & Business Media.
11. El-Demirdash, M. (2011). *The green pyramid rating system* (Vol. 1). Housing and Building National Research Centre.
12. Dou, X., Ren, F., Nguyen, M. Q., Ahamed, A., Yin, K., Chan, W. P., & Chang, V. W.-C. (2017). Review of MSWI bottom ash utilization from perspectives of collective characterization, treatment and existing application. *Renewable and Sustainable Energy Reviews, 79*, 24–38. https://doi.org/10.1016/j.rser.2017.04.020
13. del Valle-Zermeño, R., Formosa, J., Chimenos, J. M., Martínez, M., & Fernández, A. I. (2013). Aggregate material formulated with MSWI bottom ash and APC fly ash for use as secondary building material. *Waste Management, 33*, 621–627. https://doi.org/10.1016/j.wasman.2012.11.027
14. Luo, H.-L., Chen, S.-H., Lin, D.-F., & Cai, X.-R. (2017). Use of incinerator bottom ash in open-graded asphalt concrete. *Construction and Building Materials, 149*, 497–506. https://doi.org/10.1016/j.conbuildmat.2017.05.057
15. Robinson, J., Kumari, N., Srivastava, V. K., Taskaeva, N., & Mohan, C. (2022). Sustainable and environmental friendly energy materials. *Materials Today Proceedings, 69*, 494–498. https://doi.org/10.1016/j.matpr.2022.03.106
16. Nadeem, M. W., Goh, H. G., Hussain, M., Ponnusamy, V. a/p, Hussain, M., & Khan, M. A. (2020). Internet of things for green building management: A survey. In *Role of IoT in green energy systems* (Chapter 7, p. 15). IGI.
17. Kapilan, N., & Vidhya, P. (2021). Challenges and issues of IoT application in heating ventilating air conditioning systems: energy conservation using IoT. In *Role of IoT in green energy systems* (Chapter 8, pp. 171–193). IGI.
18. Asmara, Y. P. (2024). *Concrete reinforcement degradation and rehabilitation: Damages, corrosion and prevention*. Springer Nature.
19. Perpiña, C., Batista, F., & Lavalle, C. (2020). An assessment of the regional potential for solar power generation in EU-28. *Energy Policy, 88*, 86–99. https://doi.org/10.1016/j.enpol.2015.07.013
20. Jae, K., Lee, H., & Koo, Y. (2020). Research on local acceptance cost of renewable energy in South Korea: A case study of photovoltaic and wind power projects. *Energy Policy, 144*, 111684. https://doi.org/10.1016/j.enpol.2020.111684

21. Fahd Amjad, L. A. S. (2020). Identification and assessment of sites for solar farm development using GIS and density-based clustering technique: A case of Pakistan. *Renewable Energy, 155*, 761–769. https://doi.org/10.1016/j.renene.2020.03.100
22. Yoomak, S., Patcharoen, T., & Ngaopitakkul, A. (2019). Performance and economic evaluation of solar rooftop systems in different regions of Thailand. *Sustainability, 11*, 6647. https://doi.org/10.3390/su11236647
23. Tseng, K. H., Chung, M. Y., Chen, L. H., et al. (2022). A study of green roof and impact on the temperature of buildings using an integrated IoT system. *Scientific Reports, 12*, 16140. https://doi.org/10.1038/s415
24. Velásquez, C., Espín, F., Castro, M. Á., & Rodríguez, F. (2024). Energy efficiency in public lighting systems friendly to the environment and protected areas. *Sustainability, 16*(12), 5113. https://doi.org/10.3390/su16125113
25. U.S. Department of Energy. (n.d.). *LED lighting.* Energy Saver. https://www.energy.gov/energysaver/led-lighting
26. Fontoynont, M. (2018). LED lighting, ultra-low-power lighting schemes for new lighting applications. *Comptes Rendus Physique, 19*, 159–168. https://doi.org/10.1016/j.crhy.2017.10.003
27. Hossain, R., Sultana, R., Patwary, M. M., Khunga, N., Sharma, P., & Shaker, S. J. (2022). Self-healing concrete for sustainable buildings: A review. *Environmental Chemistry Letters, 20*, 1265–1273. https://doi.org/10.1007/s10311-022-01494-7
28. Hassan, Q., Abdulateef, A. M., Hafedh, S. A., Alsamari, A., Abdulateef, J., Sameen, A. Z., & Jaszczur, M. (2023). Renewable energy-to-green hydrogen: A review of main resources, routes, processes, and evaluation. *International Journal of Hydrogen Energy, 48*(46), 17383–17408. https://doi.org/10.1016/j.ijhydene.2023.05.104
29. Mathivanan, D. B., Siregar, J. P., Mat Rejab, M. R., Bachtiar, D., Asmara, Y. P., & Cionita, T. (2017). The mechanical properties of alkaline treated pineapple leaf fibre to reinforce tapioca based bioplastic resin composite. *Materials Science Forum, 882*, 66–70. https://doi.org/10.4028/www.scientific.net/msf.882.66
30. Jawaid, M., & Abdul Khalil, H. P. S. (2011). Cellulosic/synthetic fibre reinforced polymer hybrid composites: A review. *Carbohydrate Polymers, 86*(1), 1–18. https://doi.org/10.1016/j.carbpol.2011.04.024
31. Abdul Khalil, H. P. S., Bhat, I. U. H., Jawaid, M., Zaidon, A., Hermawan, D., Hadi, Y. S. (2012). Bamboo fibre reinforced biocomposites: A review. f*Materials & Design, 42*, 353–368. https://doi.org/10.1016/j.matdes.2012.05.043
32. Bledzki, A., Jaszkiewicz, A., Urbaniak, M., & Walczak, D. (2012). Biocomposites in the past and in the future. *Fibres and Textiles in Eastern Europe, 96*, 15–22.
33. Benatti, A. C. B., Pattaro, A. F., Rodrigues, A. A., Xavier, M. V., Kaasi, A., Barbosa, M. I. R., Jardini, A. L., Maciel Filho, R., & Kharmandayan, P. (2019). Bioreabsorbable polymers for tissue engineering: PLA, PGA, and their copolymers. In A. M. Holban & A. M. Grumezescu (Eds.), *Materials for biomedical engineering* (pp. 83–116). Elsevier.
34. Koushik, T. M., Miller, C. M., & Antunes, E. (2022). Bone tissue engineering scaffolds: Function of multi-material hierarchically structured scaffolds. *Advanced Healthcare Materials.* https://doi.org/10.1002/adhm.202201015
35. Mullen, E., & Morris, M. A. (2021). Green nanofabrication opportunities in the semiconductor industry: A life cycle perspective. *Nanomaterials, 11*(5), 1085. https://doi.org/10.3390/nano11051085

36. Pinto-Gómez, C., Pérez-Murano, F., Bausells, J., Villanueva, L. G., & Fernández-Regúlez, M. (2020). Directed self-assembly of block copolymers for the fabrication of functional devices. *Polymers (Basel), 12,* 2432. https://doi.org/10.3390/polym12102432
37. Liddle, J. A., & Gallatin, G. M. (2016). Nanomanufacturing: A perspective. *ACS Nano, 10,* 2995–3014. https://doi.org/10.1021/acsnano.6b02074
38. Cummins, C., Lundy, R., Walsh, J. J., Ponsinet, V., Fleury, G., & Morris, M. A. (2020). Enabling future nanomanufacturing through block copolymer self-assembly: A review. *Nano Today, 35,* 100936. https://doi.org/10.1016/j.nantod.2020.100936
39. Sarangan, A. (2016). Nanofabrication. In *Fundamentals and applications of nanophotonics* (pp. 149–184). Elsevier. https://doi.org/10.1016/B978-1-78242-464-2.00004-1
40. Black, C. T. (2007). Polymer self-assembly as a novel extension to optical lithography. *ACS Nano, 1,* 147–150. https://doi.org/10.1021/nn700051f
41. Nicolay, P., Schlögl, S., Thaler, S. M., Humbert, C., & Filipitsch, B. (2023). Smart materials for greener cities: A short review. *Applied Sciences, 13*(16), 9289. https://doi.org/10.3390/app13169289
42. Tabrizikahou, A., Kuczma, M., Łasecka-Plura, M., Farsangi, E. N., Noori, M., Gardoni, P., & Li, S. (2022). Application and modelling of shape-memory alloys for structural vibration control: State-of-the-art review. *Construction and Building Materials, 342*(B), 127975.
43. Zhang, Z., & Zhang, L. (2023). Thermochromic energy-efficient windows: Fundamentals, recent advances, and perspectives. *Chemical Reviews, 123*(11), 7025–7080. https://doi.org/10.1021/acs.chemrev.2c00647
44. Mariyaiah, Sadashiva & Sheikh, M & Khan, Nouman & Kurbet, Ramesh & Gowda, T.M.Deve. (2021). A Review on Application of Shape Memory Alloys. *International Journal of Recent Technology and Engineering,* 9, 111–120.
45. Hakami, Abdullatif & Srinivasan, Sesha & Biswas, Prasanta & Krishnegowda, Ashwini & Wallen, Scott & Stefanakos, Elias. (2022). Review on thermochromic materials: development, characterization, and applications. *Journal of Coatings Technology and Research.* 19. 10.1007/s11998-021-00558-x.
46. Fernandez, C. A., Correa, M., Nguyen, M.-T., Rod, K. A., Dai, G. L., Cosimbescu, L., Rousseau, R., & Glezakou, V.-A. (2020). Progress and challenges in self-healing cementitious materials. *Journal of Materials Science, 56,* 201–230. https://doi.org/10.1007/s11041-020-03247-z
47. Hossain, M., Zhumabekova, A., Paul, S., et al. (2020). A review of 3D printing in construction and its impact on the labor market. *Sustainability, 12*(20), 8492. https://doi.org/10.3390/su12208492
48. Amran, M., Onaizi, A. M., Fediuk, R., Vatin, N. I., Rashid, R. S. M., Abdelgader, H., & Ozbakkaloglu, T. (2022). Self-healing concrete as a prospective construction material: A review. *Materials, 15,* 3214. https://doi.org/10.3390/ma15093214
49. Lu, Y. (2017). Industry 4.0: A survey on technologies, applications, and open research issues. *Journal of Industrial Information Integration, 6,* 1–10. https://doi.org/10.1016/j.jii.2017.01.002
50. Ardanza, A., Moreno, A., Segura, A., de la Cruz, M., & Aguinaga, D. (2019). Sustainable and flexible industrial human-machine interfaces to support adaptable applications in the industry 4.0 paradigm. *International Journal of Production Research, 57*(12), 4045–4059. https://doi.org/10.1080/00207543.2019.1596921

51. Arunachalam, D., Kumar, N., & Kawalek, J. P. (2018). Understanding big data analytics capabilities in supply chain management: Unravelling the issues, challenges, and implications for practice. *Transportation Research Part E: Logistics and Transportation Review, 114,* 416–436. https://doi.org/10.1016/j.tre.2018.02.002
52. Dubey, R., Gunasekaran, A., Childe, S. J., & Papaloucas, I. (2021). Industry 4.0 technologies adoption and its impact on supply chain sustainability: A structural model analysis. *Production Planning & Control, 32*(3), 231–247. https://doi.org/10.1080/09537287.2020.1817461
53. Ahuett-Garza, H., & Kurfess, T. (2018). A brief discussion on the trends of habilitating technologies for Industry 4.0 and smart manufacturing. *Manufacturing Letters, 15,* 60–63. https://doi.org/10.1016/j.mfglet.2018.05.001
54. Frank, A. G., Dalenogare, L. S., & Ayala, N. F. (2019). Industry 4.0 technologies: Implementation patterns in manufacturing companies. *International Journal of Production Economics, 210,* 15–26. https://doi.org/10.1016/j.ijpe.2019.01.014
55. Awogbemi, O., & Kallon, D. V. V. (2021, October 4–6). Impact of the fourth industrial revolution on waste biomass conversion techniques. In *Proceedings of the SAIIE32, Steps,* 352–365.
56. Massaro, M., Secinaro, S., Dal Mas, F., Brescia, V., & Calandra, D. (2021). Industry 4.0 and circular economy: An exploratory analysis of academic and practitioners' perspectives. *Business Strategy and the Environment, 30*(2), 1213–1231. https://doi.org/10.1002/bse.2679
57. Buchi, G., Cugno, M., & Castagnoli, R. (2020). Smart factory performance and industry 4.0. *Technological Forecasting and Social Change, 150,* 119790. https://doi.org/10.1016/j.techfore.2019.119790
58. Bagnoli, C., Dal Mas, F., & Massaro, M. (2019). The 4th industrial revolution: Business models and evidence from the field. *International Journal of E-Services and Mobile Applications (IJESMA), 11*(3), 34–47.
59. Yao, X., Zhou, J., Lin, Y., Li, Y., Yu, H., & Liu, Y. (2019). Smart manufacturing based on cyber-physical systems and beyond. *Journal of Intelligent Manufacturing, 30*(8), 2805–2817. https://doi.org/10.1007/s10845-019-01468-5
60. Umar, M., Khan, S. A. R., Yusliza, M. Y., Ali, S., & Yu, Z. (2021). Industry 4.0 and green supply chain practices: An empirical study. *International Journal of Productivity and Performance Management. IDEAS, 12.0633*
61. Frank, A. G., Dalenogare, L. S., & Ayala, N. F. (2019). Industry 4.0 technologies: Implementation patterns in manufacturing companies. *International Journal of Production Economics, 210,* 15–26. https://doi.org/10.1016/j.ijpe.2019.02.003
62. Umar, M., Khan, S., Zia-ul-haq, H. M., Yusliza, M. Y., & Farooq, K. (2021). The role of emerging technologies in implementing green practices to achieve sustainable operations. *The TQM Journal.* https://doi.org/10.1108/TQM-06-2021-0172
63. Islam, M. J., Roy, S. K., Miah, M., & Das, S. K. (2020). A review on corporate environmental reporting (CER): An emerging issue in the corporate world. *Canadian Journal of Business and Information Studies, 2*(3), 45–53.
64. Abdulla, Y., Ebrahim, R., & Kumaraswamy, S. (2020). *Artificial intelligence in the banking sector: Evidence from Bahrain.* 2020 International Conference on Data Analytics for Business and Industry: Way Towards a Sustainable Economy (ICDABI). 1-6. https://doi.org/10.1109/ICDABI51230.2020.9325600

65. Jayadatta, S., & Nitin, S. (2017). Opportunities, challenges, initiatives, and avenues for green banking in India. *International Journal of Business and Management Invention, 6*(2), 10–15.
66. Zaheer, A., & Singh, A. (2023). Review on impact of AI on green banking towards sustainability with special reference to Indian public and private banks. *Journal of Survey in Fisheries Sciences, 10*(4S), 2475–2492.
67. Market.us. (2024). *Global green finance market size, trends, and forecast to 2033.* https://www.market.us/report/green-finance-market/
68. He, L., Zhang, L., Zhong, Z., Wang, D., & Wang, F. (2019). Green credit, renewable energy investment and green economy development: Empirical analysis based on 150 listed companies of China. *Journal of Cleaner Production, 208,* 363–372. https://doi.org/10.1016/j.jclepro.2018.10.119
69. Cigu, E. (2020). The role of green finance: An overview. In M. Tofan, I. Bilan, & E. Cigu (Eds.), *European Union financial regulation and administrative area– EUFIRE* (pp. 657–669). Alexandru Ioan Cuza University Publishing House.
70. Alharbi, S. S., Al Mamun, M., Boubaker, S., & Rizvi, S. K. A. (2023). Green finance and renewable energy: Worldwide evidence. *Energy Economics, 118,* 106499. https://doi.org/10.1016/j.eneco.2022.106499
71. Campiglio, E. (2016). Beyond carbon pricing: The role of banking and monetary policy in financing the transition to a low-carbon economy. *Ecological Economics, 121,* 220–230. https://doi.org/10.1016/j.ecolecon.2015.03.020
72. Li, Z., Kuo, T.-H., Siao-Yun, W., & The Vinh, L. (2022). Role of green finance, volatility, and risk in promoting investments in renewable energy resources in the post-COVID-19 era. *Resources Policy, 76,* 102563. https://doi.org/10.1016/j.resourpol.2022.102563
73. Lee, J. W. (2020). Green finance and sustainable development goals: The case of China. *Journal of Asian Finance, Economics, and Business, 7*(7), 577–586. https://doi.org/10.13106/jafeb.2020.vol7.no7.577

Index

For Product Safety Concerns and Information please contact our EU
representative GPSR@taylorandfrancis.com
Taylor & Francis Verlag GmbH, Kaufingerstraße 24, 80331 München, Germany

www.ingramcontent.com/pod-product-compliance
Ingram Content Group UK Ltd.
Pitfield, Milton Keynes, MK11 3LW, UK
UKHW022124190926
475702UK00013B/1640